4.00
N/M SALE

From The Library of
PATRICK J. HOGAN M.D.
2121 Addison
HOUSTON, TEXAS 77030

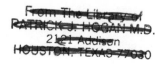
From The Library of
PATRICK J. HOGAN M.D.
2121 Addison
HOUSTON, TEXAS 77030

# Beat the Odds

## Microcomputer Simulations of Casino Games

D1276085

# The Hayden Microcomputer Series

*Consulting Editor: Ted Lewis, Oregon State University*

†Consulting Editor: Sol Libes, Amateur Computer Group of New Jersey and
Union Technical Institute

# Beat the Odds

## Microcomputer Simulations of Casino Games

**HANS SAGAN**

*Professor of Mathematics*
*North Carolina State University*

HAYDEN BOOK COMPANY, INC.
Rochelle Park, New Jersey

**Library of Congress Cataloging in Publication Data**

Sagan, Hans.
   Beat the odds.

   (The Hayden microcomputer series)
   Bibliography: p.
   Includes index.
   1. Games of chance (Mathematics)—Data processing.
2. Gambling—Data processing. 3. Digital computer
simulation. 4. Microcomputers. I. Title.
II. Series: Hayden microcomputer series.
QA271.S23       795'.028'54      80-20093

ISBN 0-8104-5181-6

*Copyright © 1980 by HAYDEN BOOK COMPANY, INC.* All rights reserved.
No part of this book may be reprinted, or reproduced, or utilized in any
form or by any electronic, mechanical, or other means, now known or
hereafter invented, including photocopying and recording, or in any information storage and retrieval system, without permission in writing from
the Publisher.

*Printed in the United States of America*

| 3 | 4 | 5 | 6 | 7 | 8 | 9 | PRINTING |
|---|---|---|---|---|---|---|---|

| 81 | 82 | 83 | 84 | 85 | 86 | 87 | 88 | YEAR |
|----|----|----|----|----|----|----|----|------|

# Preface

To gamble is to take a risk in the hope of gaining an advantage. Wouldn't we all like to get something for next to nothing with minimal effort? Rare is the saint indeed who has never succumbed to the temptation of games of chance that hold out the promise of untold riches for a token investment. Whether we flip for coffee, play bingo with religious overtones in our church's basement recreation room, play the numbers or ponies with the friendly neighborhood (the accent is on "hood") bookie, participate in the office basketball or football pool, or play baccarat in Monte Carlo for sums few of us can grasp conceptually, we are responding to one of man's most elemental motivating forces: a rapacious desire to amass secular goods.

It is not the purpose of this volume to promote gambling or to entice anybody who has not succumbed already. To be sure, gambling is immoral and, in most states of the union, against the law. Besides, you can't win in the long run anyway. Of course, there are winners, even big winners, but since casinos are not in business for their health but for a healthy rake-off, the money has to come from us, the hundreds of thousands of little losers and big losers. And what about the winners? Do they have the strength to stay away from the tables, or will they be drawn back to the pits and blow it all, winding up losers like the rest of us?

In this book, we try to explain how to play certain casino games, make an attempt to convince you that, in the long run, you cannot win (except, possibly, at blackjack), explain some popular systems and their pitfalls (before you fall into the pit), and, above all, provide very realistic computer simulations of the games themselves to enable you to "gamble" to your heart's content without ever losing any real money. If, after all that, you are still inclined to plunge ahead and take on the immutable laws of mathematics, don't blame us if things go awry.

For the purpose of our demonstration, we have selected five casino games: *trente-et-quarante* (thirty-and-forty), *roulette*, *chemin-de-fer*, *craps*, and *blackjack*.

All games of chance, when stripped of their external trappings, boil down to betting on a chance event. Such an event, usually characterized by a number, is generated by some random device. This device may be one or more well-shuffled decks of cards such as in trente-et-quarante, chemin-de-fer, and blackjack, or it may be a mechanical device such as a roulette wheel or a pair of dice. For variety's sake, our presentation will alternate between games that use cards and games that use mechanical devices.

We are going to start out with trente-et-quarante for two reasons: It is a card game that contains all the elements of the more complicated card games, chemin-de-fer and blackjack, in rudimentary form. These elements are shuffling, dealing, and counting. Second, the betting involved is similar to, but simpler than, the betting on even chances offered by roulette. (Trente-et-quarante also affords slightly more favorable odds than roulette.)

Roulette, probably the most widely known casino game, comes next. Here, the chance events—numbers—are generated by a mechanical device: a wheel. Roulette as played in European and South African casinos differs somewhat from the game played in Nevada and Atlantic City—the only places in the United States where it is legal. European roulette affords much better odds than American roulette. We deal with both versions, and our computer program simulates both games.

We then turn to a discussion of chemin-de-fer, a variation of baccarat. We have chosen it instead of baccarat because it is more easily adaptable to a computer simulation and is also, quite frankly, the more interesting game. The American version of baccarat is so similar to chemin-de-fer that the cognoscenti should not encounter any difficulties in converting our chemin-de-fer program into an American baccarat program.

For a change of pace, we turn next to craps. Here, the random event is generated by the rolling of two dice. As does roulette, craps offers a wide variety of bets, ranging from ones affording quite reasonable odds all the way to the worst sucker come-ons in gambling history.

Finally, we discuss blackjack, the one casino game where the player may actually have a chance. In all the other games discussed, the player need not know anything beyond the most rudimentary rules about placing bets. In blackjack, the player has to know a great deal about strategy if he does not want to throw his money away. Our computer program, in conjuction with our instructions, ought to go a long way towards teaching the reader to play a good game of blackjack.

If he follows our advice and studies some of the literature on more advanced strategies, he may wind up playing a very good game of blackjack.

Assuming that "the mob" has not tampered with the roulette wheel, the dice, and the cards, and that the dealers and attendants are honest (which cannot be taken for granted; see the Bibliography, Ref. [13], p. 141ff), probability theory enables us to make certain predictions about winning, losing, and the house-take. The most rudimentary maxims of probability theory are summarized in Appendix A. A knowledge of this material is not required for an understanding of the main text provided that the reader is willing to accept our figures pertaining to expected losses on faith. For the convenience of the reader, we have classified most of the admissible bets in ascending order of expected losses in Appendix C.

All our programs allow the operator to talk to the computer in plain English (or French, as the case may be) and make it respond in like manner. We have never been able to warm up to computer games where one has to enter 37 for "yes," 49.7 for "no," 63 for "green," 131 for "my sister-in-law is pregnant," and such. We have also arranged for the computer to do all the bookkeeping. This feature is very important if one wants to measure how well (or how poorly) one is doing versus the house or other players. Finally, we have added a lot of frills to make the simulations as realistic as possible. All of this has been accomplished with very few elementary programming tools and techniques. The first several chapters of almost any introduction to BASIC should suffice to prepare the reader for the programs in this volume (see, for example, Ref. [8]).

Central to all our programs is the RND (random) function. If you do not have an adequate pseudo-random-number generator, you'll have to program your own and incorporate it into our programs as a sub-routine (see also Ref. [8], p. 168). In addition to the RND-function, we have made extensive use of the INPUT statement (for numerical values as well as strings), of the sub-string capability of the HP 2000 ACCESS BASIC (the language we are working in), of FOR-NEXT loops, of the GOTO and computed GOTO statement, and of the GOSUB statement. The availability of the logical operators AND and OR was an enormous help (but not indispensable) when programming the computer to check on the outcome of a game or to follow a certain strategy. (An AND may always be replaced by two IF-THEN statements in series and an OR by two parallel IF-THEN statements.) Extensive use was made of the INT-function, and some use of the ABS-function and the SGN-function.

Other than these aids, we made use of the PRINT, IF-THEN statements, and, for the purpose of formatting the output, the LIN- and

TAB-functions and the PRINT USING and IMAGE statements. In some strategic places, MIN was used to advantage. In many instances, we had to use two or more lines in our programs where one would normally have sufficed in order to avoid having any of them exceed 63 characters in the programs and in the print-out. The restriction to 63 characters has nothing to do with our computer or terminal but everything to do with the trim size of this book and our unwillingness to have—in the interest of readability—the print-outs reduced to less than 70 percent of their original size. This led, on occasion, to some awkward programming. In particular, lengthy conjunctions had to be negated and then, using DeMorgan's law, replaced with a disjunction of negations that, in turn, could be distributed over many lines.

The versatility of our programs (the program Roulette, for example, accommodates the placing of 32 different kinds of bets, each in two languages) led to greater complexity ("knotting") than one would ordinarily like. Some of it could have been avoided by increasing the length of the programs, which is, unfortunately, another undesirable feature. To make the programs more readable, we made liberal use of REM statements. They are easily spotted; their line numbers do not end with a zero as all the others do. The use of two-dimensional arrays would have led to a number of simplifications. We resisted the temptation and avoided them because they are not yet widely available. For some hints on the discrepancies between major BASIC dialects, see Appendix B.

The international language of gambling is French, and at least some French terms and phrases are used in most casinos outside the United States. To prepare our reader for the eventuality of finding himself in foreign surroundings, we have incorporated the most commonly used French expressions into our games. Do not be put off by the occasional French word or sentence. We use a total of only 99 French words—a vocabulary that should not place an undue burden on the reader. Watch your step, however. Before you move into the glamorous world of international gambling, ask your more cosmopolitan friends about the pronunciation or you may run into some problems. For the convenience of our readers, we have supplied a French-English mini-dictionary in Appendix D.

Each of the five chapters has essentially the same structure. It begins with a computer run that displays as many facets of the program as practical, followed by an explanation of the objectives and the physical execution of the game. After that comes a discussion of the various bets that are acceptable and how to place them. Some systems and/or strategies are laid out next, and then, when the reader may be assumed to be thoroughly familiar with the game, the computer program is developed. A discussion of various modifications of the program concludes the chapter.

Our unpretentious Bibliography at the back of the book contains some readily available and reasonably well known books on gambling, programming, and loosely related topics. References to sources that are listed in the Bibliography are made by the appropriate reference-number in brackets. For example, (Ref. [12], p. 25) refers to page 25 of the work listed as Reference 12 in the Bibliography.

I wish to thank Glidrose Publications Limited, London, for their gracious permission to use several of Ian Fleming's characters in the simulated chemin-de-fer game in Chap. 3, and United Features Syndicate, who, at the behest of Charles M. Schulz, allowed the image of flamboyant "Blackjack Snoopy, the World Famous Riverboat Gambler" to occupy a pre-eminent position in Chap. 5. The connoisseur may wonder what happened to the Damon Runyon characters of "Guys and Dolls" fame, who would have been a natural for the simulated craps game in Chap. 4. Well, we asked the holders of the copyright for their permission. No dice!

Finally, I wish to thank the editors and staff of Hayden Book Company for their encouragement, their valuable assistance, and their generosity in providing computer time, and Sharon Letovsky for transliterating my hieroglyphics into eminently legible typescript.

HANS SAGAN

# Contents

*Grau, teurer Freund, ist alle Theorie . . .* [1]

---

[1] "My worthy friend, grey are all theories . . ." (Mephistopheles to student, *Faust*, Part I, by Johann Wolfgang von Goethe, translated by B. Taylor)

# Chapter 1

# trente-et-quarante

```
********************
*TRENTE-ET-QUARANTE*
********************
```

WHEN ASKED TO MAKE YOUR BETS (MESSIEURS, FAITES VOS JEUX),
ENTER ONE OF THE FOLLOWING:

```
                    ROUGE (RED)
                    NOIR (BLACK)
                    COULEUR (COLOR)
                    INVERSE (INVERSE)
                    ROUGE-COULEUR
                    ROUGE-INVERSE
                    NOIR-COULEUR
                    NOIR-INVERSE
```

HOW MUCH MONEY DO YOU HAVE ?10000

```
        *** STAND BY - THE CARDS ARE BEING SHUFFLED ***

    *** PLEASE CUT - BY ENTERING A NUMBER BETWEEN 1 AND 312 ***

?31
                *** MESSIEURS, FAITES VOS JEUX...***

?ROUGE-INVERSE

HOW MUCH DO YOU WANT TO BET ?500

            *** LE JEU EST FAIT, RIEN NE VA PLUS...***
```

Fig. 1.1 Run of the program "Trente-et-Quarante"

1

```
*** UPPER ROW ***
TEN    OF HEARTS
NINE   OF DIAMONDS
TWO    OF DIAMONDS
NINE   OF SPADES
FIVE   OF CLUBS
***    5      ***
```

```
                                    *** LOWER ROW ***
                                    EIGHT OF CLUBS
                                    SEVEN OF HEARTS
                                    KING  OF DIAMONDS
                                    NINE  OF HEARTS
                                    ***    4       ***
```

```
            *** ROUGE GAGNE,ET COULEUR       ***
```

```
            *** COUP NEUTRE...HERE WE GO AGAIN ***
```

```
*** UPPER ROW ***
THREE OF SPADES
ACE    OF DIAMONDS
KING   OF CLUBS
EIGHT OF SPADES
FIVE   OF CLUBS
FOUR   OF SPADES
***    1      ***
```

```
                                    *** LOWER ROW ***
                                    SIX    OF DIAMONDS
                                    ACE    OF CLUBS
                                    THREE OF HEARTS
                                    QUEEN OF CLUBS
                                    JACK   OF HEARTS
                                    QUEEN OF SPADES
                                    **   QUARANTE    **
```

```
            *** ROUGE PERD   ET COULEUR GAGNE ***
```

```
YOU JUST LOST  500    LOUIS !
YOU NOW HAVE  9500      LOUIS. WANT TO TRY IT AGAIN ?YES
```

```
            *** MESSIEURS, FAITES VOS JEUX...***
```

```
?ROUGE
```

```
HOW MUCH DO YOU WANT TO BET ?1000
```

```
            *** LE JEU EST FAIT, RIEN NE VA PLUS...***
```

```
*** UPPER ROW ***
TEN    OF CLUBS
EIGHT OF DIAMONDS
EIGHT OF CLUBS
```

Fig. 1.1 Run of the program "Trente-et-Quarante" (cont'd)

```
NINE  OF SPADES
***   5     ***
                                    *** LOWER ROW ***
                                    SEVEN OF DIAMONDS
                                    ACE   OF CLUBS
                                    TWO   OF DIAMONDS
                                    TEN   OF SPADES
                                    FIVE  OF DIAMONDS
                                    SIX   OF HEARTS
                                    ***   1     ***


            *** ROUGE GAGNE ET COULEUR PERD ***


YOU JUST WON  1000    LOUIS !
YOU NOW HAVE  10500   LOUIS. WANT TO TRY IT AGAIN ?YES


            *** MESSIEURS, FAITES VOS JEUX...***


?NOIR-COULEUR


HOW MUCH DO YOU WANT TO BET ?11000
YOU DON'T HAVE ENOUGH MONEY TO COVER THAT BET !
HOW MUCH DO YOU WANT TO BET ?500


            *** LE JEU EST FAIT, RIEN NE VA PLUS...***


*** UPPER ROW ***
FOUR  OF DIAMONDS
TWO   OF CLUBS
QUEEN OF DIAMONDS
EIGHT OF DIAMONDS
NINE  OF HEARTS
***   3     ***
                                    *** LOWER ROW ***
                                    KING  OF DIAMONDS
                                    SEVEN OF HEARTS
                                    FOUR  OF HEARTS
                                    TWO   OF SPADES
                                    QUEEN OF CLUBS
                                    *  3     APRES *


            *** COUP NUL ***


            *** MESSIEURS, FAITES VOS JEUX...***


?COULEUR


HOW MUCH DO YOU WANT TO BET ?10500


            *** LE JEU EST FAIT, RIEN NE VA PLUS...***
```

Fig. 1.1 Run of the program "Trente-et-Quarante" (cont'd)

```
*** UPPER ROW ***
NINE   OF CLUBS
QUEEN OF DIAMONDS
SEVEN OF SPADES
FIVE   OF DIAMONDS
***    1      ***
```

```
                              *** LOWER ROW ***
                              EIGHT OF DIAMONDS
                              SIX    OF DIAMONDS
                              SEVEN OF SPADES
                              KING   OF CLUBS
                              *   UN  APRES   *
```

```
              *** REFAIT ***
```

```
DO YOU WANT HALF YOUR STAKE BACK ?NO
```

```
         *** EN PRISON...HERE WE GO AGAIN ***
```

```
*** UPPER ROW ***
QUEEN OF HEARTS
ACE    OF SPADES
SIX    OF DIAMONDS
JACK   OF CLUBS
FIVE   OF HEARTS
***    2      ***
```

```
                              *** LOWER ROW ***
                              FOUR   OF DIAMONDS
                              THREE OF DIAMONDS
                              KING   OF HEARTS
                              KING   OF SPADES
                              JACK   OF DIAMONDS
                              ***    7      ***
```

```
       *** ROUGE PERD ,ET COULEUR        ***
```

```
YOU JUST LOST  10500     LOUIS !
```

```
YOU LOST ALL YOUR MONEY. BUZZ OFF  !
```

```
DONE
```

Fig. 1.1 Run of the program "Trente-et-Quarante" (cont'd)

## HOW THE GAME IS PLAYED

Trente-et-quarante (thirty-and-forty, a predominantly European casino game) is played at a table as depicted in Fig. 1.2. The table is covered with a green cloth subdivided by yellow lines. One diamond is black and the other one red. The *tailleur* (dealer) sits at the place

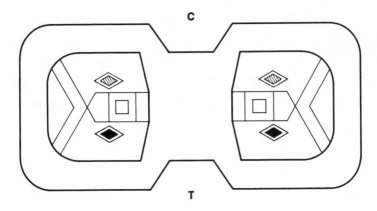

Fig. 1.2 Trente-et-quarante table

marked T in Fig. 1.2, and a *croupier* (attendant) who rakes in losses and pays out winnings sits at C.

The game is played with six decks of (ordinary) playing cards (bridge cards). Each card counts its face value (aces count 1, deuces count 2, treys count 3, etc . . . ), and the face cards (court cards) count 10 each. The *tailleur* breaks the seals of six packs, shuffles each pack individually, and then all together. A player is asked to cut, the cards are placed in a *sabot* (shoe), and the game commences with the announcement

★★★ *Messieurs, faites vos jeux* ★★★

(Gentlemen, place your bets.) Bets are placed by putting chips on the field representing the chance one wants to take. For example, a bet on *rouge* (red) is made by placing one's chip(s) on the field marked by the red diamond on the table. As soon as it looks as if all bets have been placed, the *tailleur* announces

★★★ *Le jeu est fait, rien ne va plus* ★★★

(the betting is over, no more bets) and begins dealing cards in two rows. He stops putting cards into the first row as soon as the count exceeds 30 and announces the amount by which it exceeds 30. For example, the row shown in Fig. 1.3 represents a count of $10 + 10 + 6 + 1 + 8$, or 35, and is, therefore, complete. The dealer will announce *cinq* (five), the count being 5 above 30. The maximum count of a row is, of course, 40, and the minimum count is 31 (hence the name of the game). As soon as the first row is complete, a second row is dealt until the count again exceeds 30. Again, the amount by which it exceeds 30 is announced. The only exception to this convention is when the

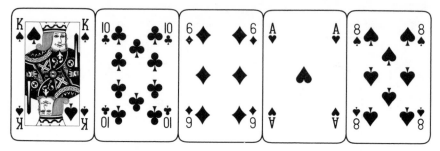

*Fig. 1.3 Completed row of cards*

count is 40. In that case, the *tailleur* announces *quarante* (forty) rather than *dix* (ten).

Once the deal is complete, there are a number of different winning possibilities:

*Noir:* If the count of the first row is lower than the count of the second row, then *noir* (black) wins.

*Rouge:* If the count of the first row is higher than the count of the second row, then *rouge* (red) wins.

*Couleur:* If *noir* wins and the first card in the first row is black, or if *rouge* wins and the first card of the first row is red, then *couleur* (color) wins.

*Inverse:* If *noir* wins and the first card of the first row is red or if *rouge* wins and the first card of the first row is black, then *inverse* (opposite) wins.

*Coup Nul* and *Refait:* If the count in both rows is the same, the dealer says *après* (after) after having announced the count of the second row (for example, if the count in both rows was 37, the dealer announces *sept après* after having dealt the second row).

If the count (of both rows) is anything else than 31, the *coup* (turn) does not count and the *tailleur* announces *coup nul* (void turn). Players may leave their stakes in place and the cards are dealt again.

If the count of both rows is 31, the *refait* (recovery) occurs. After having dealt the second row, the dealer announces *un après*, and the stakes on *rouge*, *noir*, *couleur*, and *inverse* may be put en *prison* (in prison) if the player wishes while other stakes are raked in (see the next section).

After having announced the count (actually, the excess over 30) of the second row as explained above, the *tailleur* announces the result of the *coup*, but not, as you might think, in a straightforward manner. This would be too easy for the English speaking clientele. For some unfathomable reason, black and inverse are never mentioned.

**Table 1.1    Announcement of results**

| Winning coup | Announced as |
|---|---|
| NOIR-COULEUR (black-color) | ROUGE PERD ET COULEUR GAGNE (red loses and color-wins) |
| NOIR-INVERSE (black-opposite) | ROUGE PERD ET COULEUR (red loses and so does color) |
| ROUGE-INVERSE (red-opposite) | ROUGE GAGNE ET COULEUR PERD (red wins and color loses) |
| ROUGE-COULEUR (red-color) | ROUGE GAGNE ET COULEUR (red wins and so does color) |

Table 1.1 lists the four possible results and the corresponding formula by which they are announced.

When the dealer runs out of cards (possibly in the middle of a deal), he gathers up all the cards, pompously announces

★★★ *Les cartes passent* ★★★

(the cards are gone), shuffles, has them cut, and the game goes on—as long as there are players.

## WHAT TO BET AND HOW TO GO ABOUT IT

One may bet on any of the four *chances simples* (even chances)—that is, *noir, rouge, couleur,* or *inverse*—or one may bet *à cheval* (astride) on a combination of two nonexclusive even chances.

### Betting on Chances Simples (Even Chances)

*Pay-off:*      Even money
*Odds for winning:*      3210:3379
*Expected loss:*      12.82 louis[1] per 1000 one-louis bets

To bet on *noir, rouge, couleur,* or *inverse,* you place your chip(s) on the appropriate field. A bet on *noir,* for example, is placed on the field that is labeled by the black diamond and a bet on *rouge* on the field labeled by the red diamond. Bets on *couleur* or *inverse* are placed on the remaining two spaces, as indicated in Fig. 1.4.

If you win a *coup* (turn) with one of these *chances simples,* you win even money; if you lose, your stake is raked in by the *croupier.*

In case of a *coup nul* (invalid turn, in which both rows show the same count, but *not* 31), you may let your stake ride for the next turn, you may put it on another chance, or you may withdraw it.

---

[1] There hasn't been such a (20 franc gold) coin in over 50 years, but many croupiers still hang on to that term, just as our British cousins still use *guinea* for the sum of one pound and one shilling.

In case of a *refait* (recovery, in which both rows show the count 31), you are given the choice of forfeiting half your stake or having your stake put *en prison* (in prison). In the latter case, your chip(s) is (are) moved from *rouge* or *noir* to the appropriate diamond, from *couleur* to the rectangle within the pentagon reserved for *couleur*, and from *inverse* into the inverted V (see also Fig. 1.4). If the next coup results again in a *refait*, or if you lose the next *coup*, then your stake is lost. If you win the next *coup*, then you may retrieve your stake (you break even).

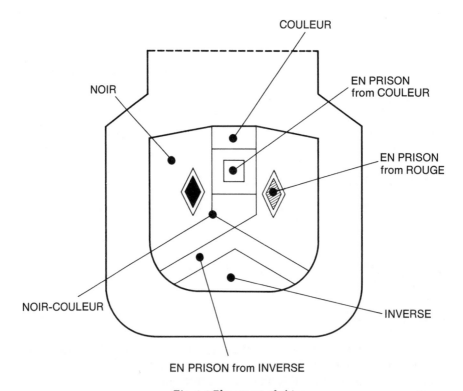

COULEUR

EN PRISON
from COULEUR

NOIR

EN PRISON
from ROUGE

NOIR-COULEUR

INVERSE

EN PRISON from INVERSE

*Fig. 1.4 Placement of chips*

There are a number of variations to this prison option. A more detailed discussion may be found in Chap. 2 on roulette.

From combinatorial arguments and elementary probabilistic considerations (see Appendix A, sections 1, 5, and 6), one obtains

$$-\frac{1}{2} \cdot \frac{169}{6589} = -0.01282$$

for the expected value of one *coup* on a one-louis bet (169 out of 6589 valid coups lead to a *refait*). The expected loss listed at the beginning of this section is based on this figure (see also Appendix A, sections 7 and 8). Consequently, the take of the house is 1.282 percent of all stakes. This payoff is better than a player can expect at roulette, where the take of the house is at least 1.35 percent (see Chap. 2).

### Betting À Cheval (Split Bet)

*Pay-off:*     Even money
*Odds for winning vs. losing:*     1605:1774
*Expected loss:*     25.65 louis per 1000 one-louis bets

You may bet *à cheval* (astride) on the following four combinations

|   |   |
|---|---|
| Noir– Couleur | Rouge– Couleur |
| Noir– Inverse | Rouge– Inverse |

by placing your chip(s) on the line separating the two fields reserved for those two chances. In Fig. 1.4 we have indicated how to place a bet on *noir-couleur*. You win if both win, lose if both lose or if there is a *refait*, and break even in case of a *coup neutre* (undecided turn), in which one wins and the other loses. Since winnings cancel losses, since nothing results from a *coup neutre*, and since you lose in case of a *refait*, the expected value of one *coup* on a one-louis bet is simply

$$-\frac{169}{6589} = -0.02565$$

which implies an expected loss as stated above (see also Appendix A, sections 5 through 8).

### PLAYING A SYSTEM

All systems that have been dreamed up for betting on an even chance at roulette apply to trente-et-quarante as well. We will discuss some of the more popularly known systems in Chap. 2 in conjunction with our explanation of roulette.

### THE COMPUTER SHUFFLES A DECK OF CARDS

We will now explain how a computer can produce a well-shuffled deck of cards. The contents of this section apply not only to trente-et-quarante, but also to chemin de fer (Chap. 3), blackjack (Chap. 5), and any other card game, for that matter (see, for example, Ref. [8], p. 147).

First of all, we have to identify the individual cards of a deck of playing cards by numbers, a job that may be accomplished in any

**Table 1.2  Numbering of cards**

|        | Club | Diamond | Heart | Spade |
|--------|------|---------|-------|-------|
| Ace    | 1    | 14      | 27    | 40    |
| 2      | 2    | 15      | 28    | 41    |
| 3      | 3    | 16      | 29    | 42    |
| 4      | 4    | 17      | 30    | 43    |
| 5      | 5    | 18      | 31    | 44    |
| 6      | 6    | 19      | 32    | 45    |
| 7      | 7    | 20      | 33    | 46    |
| 8      | 8    | 21      | 34    | 47    |
| 9      | 9    | 22      | 35    | 48    |
| 10     | 10   | 23      | 36    | 49    |
| Jack   | 11   | 24      | 37    | 50    |
| Queen  | 12   | 25      | 38    | 51    |
| King   | 13   | 26      | 39    | 52    |

number of ways. We have chosen the self-explanatory numbering displayed in Table 1.2.

If S(J) denotes the identification number of the card in position J, that is, the J-th card from the top of the deck, then

```
10   DIM S[52]
20   FOR J=1 TO 52
30   S[J]=J
40   NEXT J
```

will give us a well-ordered deck of cards, starting with the ace of clubs and ending with the king of spades.

We "shuffle" the deck as follows: We pick a number J between 1 and 52 at random and interchange the card that is on the bottom of the deck (in position 52) with the one in position J. Next, we pick a number J between 1 and 51 at random and interchange the card S(51) with the card S(J), after which we pick a number between 1 and 50 at random, and so forth. After 51 such steps, we have a well-shuffled deck of cards—or rather, the computer has. We simulate this process as follows:

```
50   FOR I=52 TO 2 STEP -1
60   J=INT(RND(1)*I)+1
70   T=S[J]
80   S[J]=S[I]
90   S[I]=T
100  NEXT I
```

Note how we have to "preserve" the old S(J) in line 70 so that we can put it into position I in line 90.

At this point, the reader might wish to check up on what happens when we add to the program the lines

```
110   FOR J=1 TO 52
120   PRINT S[J];
130   NEXT J
140   END
```

and run it several times. Here are two such runs:

**RUN**

| 48 | 25 | 23 | 19 | 33 | 43 | 32 | 22 | 36 | 37 | 49 | 10 |
| 46 | 17 | 41 | 24 | 29 | 50 | 47 | 4  | 34 | 6  | 27 | 51 |
| 52 | 3  | 38 | 15 | 44 | 35 | 20 | 2  | 16 | 9  | 11 | 5  |
| 30 | 13 | 26 | 8  | 7  | 14 | 21 | 45 | 12 | 40 | 28 | 18 |
| 1  | 42 | 39 | 31 |    |    |    |    |    |    |    |    |

**DONE**

**RUN**

| 37 | 35 | 45 | 36 | 12 | 28 | 11 | 32 | 33 | 30 | 4  | 42 |
| 3  | 52 | 50 | 17 | 31 | 20 | 34 | 6  | 41 | 9  | 14 | 38 |
| 19 | 29 | 21 | 44 | 40 | 16 | 43 | 46 | 23 | 22 | 7  | 26 |
| 1  | 8  | 51 | 18 | 10 | 5  | 24 | 47 | 27 | 25 | 48 | 15 |
| 49 | 13 | 2  | 39 |    |    |    |    |    |    |    |    |

**DONE**

Now it is time to "cut the deck":

```
110   PRINT "PLEASE CUT - BY ENTERING A NUMBER BETWEEN 1 AND 52 ";
120   INPUT Z
```

We remove the top $Z$ cards and put them on the bottom of the deck. Then, the card in position $Z+1$ moves to position 1, the card in position $Z+2$ moves into position 2, and so forth. In general, the card in position $Z+J$ moves into position $J$, and the last card will move into position $52-Z$. The card that was originally in position 1 will move into position $52-Z+1$, the card in second position moves into position $52-Z+2$, and so forth, until the card in position $Z$ moves into position $52-Z+Z(= 52)$, that is, into the last position.

If $T(J)$ denotes the identification number of the card in position $J$ *after* the cut, then, by what we said above,

```
130   FOR J=1 TO 52
140   T[J]=S[J+Z-52*INT((J+Z-1)/52)]
150   NEXT J
```

[Note that $INT((J+Z-1)/52)$ equals 0 for $J+Z \le 52$ and equals 1 for $52 < J+Z \le 104$ so that, in effect, the argument of S never exceeds 52.]

Now that our deck is shuffled and cut, the reader might wish to check it again by asking for a print-out. If you do, do not forget to add the lines

```
 10   DIM S(52), T(52)
160   END
```

Finally, we have to convert the identification numbers back into cards again by making the computer translate, for example, the number 35 into the nine of hearts.

For some card games, the suit a card belongs to is quite immaterial. If that is so, then all we have to obtain is each card's "value." Fortunately, just three lines,

```
160   FOR J=1 TO 52
170   VCJ]=TCJ]-13*INT((TCJ]-1)/13)
180   NEXT J
```

will reduce the identification numbers modulo 13 (see Table 1.2), and these are, in essence, each card's value, with the understanding that a 1 is an ace, an 11 is a jack, a 12 is a queen, and a 13 is a king.

Here is a very primitive program for card identification:

```
190   V$="123456789TJQK"
200   FOR J=1 TO 52
210   PRINT V$CVCJ],VCJ]]" ";
220   NEXT J
```

It will, in conjunction with the preceding instructions and the lines

```
 10   DIM S(52),T(52),V(52),V$(13)
230   END
```

print out the values of a shuffled and cut deck. Note that T stands for "ten." Here is a run of the program:

```
RUN


PLEASE CUT - BY ENTERING A NUMBER BETWEEN 1 AND 52 ?13
1 Q K 3 8 2 1 6 4 J 1 4 T 6 9 T K 9 9 3 2 6 J 8 Q 4 1 6 K 8 8 T 5 7 5 2
5 5 3 Q 2 T 7 4 K 9 J 7 Q 7 J 3
DONE
```

In trente-et-quarante it is important whether the first card in the first row is red or black. Even in cases where the suit is of no consequence, it adds atmosphere and realism and also aids the memory of the so-called "counters" (gamblers who keep mental track of some or all cards that have been played and recalculate the odds every time a new card is played) if the suit is mentioned. The instructions

```
210   LCJ]=INT((TCJ]-1)/13)+1
220   S$="CDHS"
```

assign 1 to clubs, 2 to diamonds, 3 to hearts, and 4 to spades. Hence

```
220   S$="CDHS"
230   PRINT V$[V[J],V[J]]" OF "S$[L[J],L[J]]" , ";
240   NEXT J
```

when augmented by

```
250   END
```

and the addition of L(52), S$(4) to line 10 will yield a print-out similar
to the following:

**RUN**

```
PLEASE CUT - BY ENTERING A NUMBER BETWEEN 1 AND 52 ?43
8 OF H , Q OF C , Q OF D , 4 OF D , K OF S , 9 OF D , 9 OF S , 8 OF D ,
2 OF D , Q OF H , 2 OF C , 3 OF D , 3 OF C , 7 OF H , 6 OF S , 2 OF H ,
1 OF S , 7 OF C , T OF H , 8 OF S , 8 OF C , J OF H , T OF S , 5 OF H ,
T OF D , K OF D , Q OF S , T OF C , 5 OF D , 1 OF C , 6 OF H , J OF D ,
6 OF C , 4 OF S , J OF S , 3 OF H , 7 OF S , 5 OF C , J OF C , K OF C ,
5 OF S , 1 OF H , K OF H , 3 OF S , 4 OF H , 6 OF D , 1 OF D , 4 OF C ,
9 OF H , 2 OF S , 9 OF C , 7 OF D ,
DONE
```

This does not satisfy us, however. We have decided to go first
class with the following de luxe shuffle:

```
10    DIM S[52],T[52],V[52],L[52],V$[65],S$[32]
20    PRINT "STAND BY - THE CARDS ARE BEING SHUFFLED."
30    FOR J=1 TO 52
40    S[J]=J
50    NEXT J
60    FOR I=52 TO 2 STEP -1
70    J=INT(RND(1)*I)+1
80    T=S[J]
90    S[J]=S[I]
100   S[I]=T
110   NEXT I
120   PRINT "PLEASE CUT - BY ENTERING A NUMBER BETWEEN 1 AND 52 ";
130   INPUT Z
140   FOR J=1 TO 52
150   T[J]=S[J+Z-52*INT((J+Z-1)/52)]
160   V[J]=T[J]-13*INT((T[J]-1)/13)
170   L[J]=INT((T[J]-1)/13)
180   NEXT J
190   V$="ACE    TWO    THREEFOUR FIVE SIX   SEVENEIGHTNINE TEN   JACK QUEENKING "
200   S$="CLUBS    DIAMONDSHEARTS  SPADES  "
210   FOR J=1 TO 52
220   PRINT V$[1+5*(V[J]-1),5+5*(V[J]-1)]" OF "S$[1+8*L[J],8+8*L[J]]" , ";
230   NEXT J
240   END
```

Here is a sample run:

```
STAND BY - THE CARDS ARE BEING SHUFFLED.
PLEASE CUT - BY ENTERING A NUMBER BETWEEN 1 AND 52 ?27
SIX     OF CLUBS     , TEN     OF DIAMONDS , FOUR   OF CLUBS     , SEVEN OF
CLUBS     , FIVE    OF DIAMONDS , TEN     OF HEARTS   , JACK   OF CLUBS     ,
QUEEN OF SPADES    , TEN     OF CLUBS     , QUEEN OF DIAMONDS , KING    OF
CLUBS     , ACE     OF DIAMONDS , SIX     OF SPADES    , EIGHT OF HEARTS    ,
SEVEN OF HEARTS    , FOUR    OF SPADES    , TWO     OF SPADES    , NINE    OF
CLUBS     , NINE    OF DIAMONDS , SEVEN OF SPADES    , ACE     OF HEARTS    ,
THREE OF SPADES    , JACK    OF SPADES    , FIVE    OF SPADES    , TWO     OF
CLUBS     , ACE     OF CLUBS     , THREE OF DIAMONDS , JACK    OF DIAMONDS ,
QUEEN OF HEARTS    , EIGHT OF CLUBS     , JACK    OF HEARTS    , EIGHT OF
SPADES    , QUEEN OF CLUBS     , NINE    OF SPADES    , TWO     OF DIAMONDS ,
THREE OF CLUBS     , SIX     OF DIAMONDS , TWO     OF HEARTS    , KING    OF
HEARTS    , THREE OF HEARTS    , FOUR    OF DIAMONDS , FIVE    OF HEARTS    ,
KING    OF SPADES    , NINE    OF HEARTS    , SIX     OF HEARTS    , TEN     OF
SPADES    , KING    OF DIAMONDS , FIVE    OF CLUBS     , EIGHT OF DIAMONDS ,
SEVEN OF DIAMONDS , FOUR    OF HEARTS    , ACE     OF SPADES    ,
DONE
```

Some microcomputers (such as the PET) actually flash color images of playing cards on the cathode-ray display screen when properly programmed to do so. The reader who owns such a computer or has access to one may find it worth his while to combine our shuffle with his computer's capacity for displaying actual images of cards to obtain a spectacular result.

## THE COMPUTER PROGRAM

It is not difficult to develop a program for trente-et-quarante that simulates the shuffling, dealing, and the announcement of the result of the coup—if we forego the luxuries of having the computer spell out the names and suits of the dealt cards, accept our bets, decide whether we lost or won, do our bookkeeping, and reshuffle the six decks without prompting if it runs out of cards (in this case, if no provisions for a reshuffle have been made and it does run out of cards, it will send a message such as "SUBSCRIPT OUT OF BOUNDS IN LINE . . ." or something to that effect). We will disregard, for the moment, the time-honored ritual of announcing the results and have the computer state the result of the coup in a straightforward manner instead.

First, we must make the computer shuffle the cards, each of the six decks individually (see the previous section). This is accomplished by the following program:

```
20    PRINT "STAND BY - THE CARDS ARE BEING SHUFFLED."
30    FOR W=0 TO 5
40    FOR J=1+52*W TO 52+52*W
50    S[J]=J
```

```
 60   NEXT J
 70   FOR I=52+52*W TO 2+52*W STEP -1
 80   J=INT(RND(1)*I)+1
 90   T=S[J]
100   S[J]=S[I]
110   S[I]=T
120   NEXT I
130   NEXT W
```

These instructions, to be augmented by an appropriate dimension statement in a line 10, shuffle six decks of cards individually. The cards in deck number 1 are identified by the numbers from 1 to 52, in deck number 2 by 53 to 104, and so forth.

Next, we have the computer shuffle the six decks together, make provisions for the operator to cut, and unscramble the code:

```
140   FOR I=312 TO 2 STEP -1
150   J=INT(RND(1)*I)+1
160   T=S[J]
170   S[J]=S[I]
180   S[I]=T
190   NEXT I
200   PRINT "PLEASE CUT - BY ENTERING A NUMBER BETWEEN 1 AND 312."
210   INPUT Z
220   FOR J=1 TO 312
230   T[J]=S[J+Z-312*INT((J+Z-1)/312)]
240   NEXT J
250   FOR J=1 TO 312
260   S[J]=T[J]-52*INT((T[J]-1)/52)
270   V[J]=S[J]-13*INT((S[J]-1)/13)
280   L[J]=INT((S[J]-1)/13)
```

So far, the program furnishes us with six decks, shuffled and cut, each card in position J (J = 1, 2, 3, . . . , or 312) and identified by its value V(J) (= 1, 2, 3, . . . , or 13) and suit L(J) (= 0, 1, 2, or 3).

Since all court cards count 10, we have to convert the values 11, 12, and 13 to 10 by means of the following lines:

```
290   G[J]=V[J] MIN 10
300   NEXT J
```

If eight decks of cards are desired rather than six, a simple conversion will yield the desired program. Here is what one has to do:

```
 30   FOR W=0 TO 7
140   FOR I=416 TO 2 STEP -1
200   PRINT "PLEASE CUT - BY ENTERING A NUMBER BETWEEN 1 AND 416";
220   FOR J=1 TO 416
230   T(J)=S(J+Z-416*INT((J+Z-1)/416)
250   FOR J=1 to 416
```

(For the appropriate DIM statement, see page 17.)

If S and T are used to keep track of the count in the upper and lower rows, respectively, and if we denote by K the number of cards that have been used up after completion of the first row (which includes all cards that have been played in previous deals) and by P the number of cards that have been used up after completion of the second row, and if C denotes the suit (0, 1, 2, or 3) of the first card in the first row (which is needed for a decision between *couleur* and *inverse*), then

```
310   V$="123456789TJQK"
320   S$="CDHS"
330   P=0
340   C=L[P+1]
350   PRINT "UPPER ROW:"
360   S=0
370   FOR J=P+1 TO 312
380   S=S+G[J]
390   PRINT V$[V[J],V[J]]" OF "S$[L[J]+1,L[J]+1]",";
400   IF S>30 THEN 420
410   NEXT J
420   K=J
430   PRINT "* "S-30" *"
440   PRINT "LOWER ROW:"
450   T=0
460   FOR J=K+1 TO 312
470   T=T+G[J]
480   PRINT V$[V[J],V[J]]" OF "S$[L[J]+1,L[J]+1]",";
490   IF T>30 THEN 510
500   NEXT J
510   P=J
520   PRINT "* "T-30" *"
```

will take care of the deal. The computer is now ready to announce the result of the *coup*:

```
530   IF S<T THEN 610
540   IF S=T THEN 640
550   PRINT "ROUGE, ";
560   IF C=1 OR C=2 THEN 590
570   PRINT "INVERSE"
580   GOTO 680
590   PRINT "COULEUR"
600   GOTO 680
610   PRINT "NOIR, ";
620   IF C=1 OR C=2 THEN 570
630   GOTO 590
640   IF T <> 31 THEN 670
650   PRINT "REFAIT"
660   GOTO 680
670   PRINT "COUP NUL"
680   PRINT "DO YOU WANT TO PLAY AGAIN ";
690   INPUT Y$
700   IF Y$[1,1]="Y" THEN 340
710   END
```

After we add the line

```
10   DIM S[312],T[312],V[312],L[312],G[312],V$[13],S$[4],Y$[3]
```

we are in possession of a workable program. (For eight decks, 312 must be replaced by 416 as the argument of S, T, V, L, G.) Here is a trial run:

```
STAND BY - THE CARDS ARE BEING SHUFFLED.
PLEASE CUT - BY ENTERING A NUMBER BETWEEN 1 AND 312.
?311
UPPER ROW:
1 OF S,3 OF D,6 OF H,7 OF C,Q OF D,1 OF C,2 OF S,2 OF C,*  2       *
LOWER ROW:
T OF D,7 OF S,K OF C,K OF C,*  7       *
NOIR, COULEUR
DO YOU WANT TO PLAY AGAIN ?YES
UPPER ROW:
5 OF H,9 OF D,5 OF S,3 OF H,8 OF C,1 OF D,*  1       *
LOWER ROW:
K OF D,6 OF C,T OF S,T OF C,*  6       *
NOIR, INVERSE
DO YOU WANT TO PLAY AGAIN ?YES
UPPER ROW:
Q OF S,J OF C,T OF D,4 OF C,*  4       *
LOWER ROW:
Q OF S,9 OF H,6 OF C,5 OF D,5 OF D,*  5       *
NOIR, COULEUR
DO YOU WANT TO PLAY AGAIN ?YES
UPPER ROW:
K OF D,8 OF C,5 OF D,9 OF S,*  2       *
LOWER ROW:
8 OF H,4 OF C,Q OF D,Q OF H,*  2       *
COUP NUL
DO YOU WANT TO PLAY AGAIN ?YES
UPPER ROW:
T OF D,9 OF D,1 OF H,5 OF H,8 OF D,*  3       *
LOWER ROW:
3 OF D,1 OF H,4 OF D,3 OF D,5 OF S,J OF H,4 OF H,5 OF S,*  5       *
NOIR, INVERSE
DO YOU WANT TO PLAY AGAIN ?YES
UPPER ROW:
8 OF D,8 OF D,J OF C,4 OF S,1 OF S,*  1       *
LOWER ROW:
7 OF H,7 OF H,T OF S,2 OF S,8 OF H,*  4       *
NOIR, INVERSE
DO YOU WANT TO PLAY AGAIN ?YES
UPPER ROW:
1 OF D,3 OF H,8 OF S,9 OF S,K OF H,*  1       *
LOWER ROW:
T OF H,K OF D,J OF C,7 OF H,*  7       *
NOIR, INVERSE
DO YOU WANT TO PLAY AGAIN ?YES
UPPER ROW:
T OF D,Q OF S,3 OF C,5 OF C,T OF D,*  8       *
LOWER ROW:
8 OF H,5 OF H,2 OF C,6 OF S,Q OF D,*  1       *
ROUGE, COULEUR
DO YOU WANT TO PLAY AGAIN ?NO

DONE
```

In order to obtain a more elaborate program (specifically, the one displayed in Fig. 1.5) that is capable of producing a run such as the one in Fig. 1.1, we have to add five essentially new elements:

1. A (numerical) coding of the bet
2. A (numerical) coding of the result of the *coup*
3. A print-out of the result of the *coup* as shown in Table 1.1
4. Automatic reshuffling of the cards when the computer runs out of cards
5. Provisions for a *refait* and *coup neutre*

We use the code numbers 0, 1, 2, 3, 4, 5, 6, 7 for the eight possible bets shown in Table 1.3, which we assign by means of the loop 330–350 (Fig. 1.5), which checks the fifth, sixth, and seventh characters of the input A\$ (line 310), comparing them with three consecutive characters in the test-word G\$ of line 190.

For the result of the *coup*, we use the code numbers $D = 2$ for *couleur*, $D = 3$ for *inverse*, $E = 0$ for *noir*, and $E = 1$ for *rouge*. (See lines 790, 820, 840, 880, 910, and 940 in Fig. 1.5.) These codes are then used in lines 1220–1280 to decide the outcome of the game.

**Table 1.3   Betting code**

| Bet | Code A |
|---|---|
| N O I R | 0 |
| R O U G E | 1 |
| C O U L E U R | 2 |
| I N V E R S E | 3 |
| N O I R - C O U L E U R | 4 |
| R O U G E - C O U L E U R | 5 |
| N O I R - I N V E R S E | 6 |
| R O U G E - I N V E R S E | 7 |

The print-out of the result of the *coup* is accomplished by means of the string R\$, defined in line 760, and filled out appropriately in lines 780, 810, 850, 870, 900, and 930.

Before the computer deals the first row of cards, it checks how many are left by means of the line

$$450 \quad \text{IF P} < 312 \text{ THEN } 470$$

which passes control (passively) to line 460 when $P = 312$, that is, when all cards have been used up. From line 460, control passes to the "shuffling and cutting" sub-routine 1470–1840. If the computer runs out of cards while dealing the upper row, then control will pass from line 530 to line 460 and from there to the "shuffling and cutting"

```
10    PRINT TAB(21)"*********************"
20    PRINT TAB(21)"*TRENTE-ET-QUARANTE*"
30    PRINT TAB(21)"*********************"
40    PRINT LIN(1)
50    PRINT "WHEN ASKED TO MAKE YOUR BETS (MESSIEURS, FAITES VOS ";
60    PRINT "JEUX),"
70    PRINT "ENTER ONE OF THE FOLLOWING:"
80    PRINT LIN(1)
90    PRINT TAB(23)"ROUGE (RED)"
100   PRINT TAB(23)"NOIR (BLACK)"
110   PRINT TAB(23)"COULEUR (COLOR)"
120   PRINT TAB(23)"INVERSE (INVERSE)"
130   PRINT TAB(23)"ROUGE-COULEUR"
140   PRINT TAB(23)"ROUGE-INVERSE"
150   PRINT TAB(23)"NOIR-COULEUR"
160   PRINT TAB(23)"NOIR-INVERSE"
170   DIM S[312],T[312],V[312],L[312],G[312],V$[65],S$[65]
180   DIM A$[13],R$[29],G$[24],Y$[4]
190   G$="  E  EURRSE-COE-C-INE-I"
200   V$[1,45]="ACE  TWO  THREEFOUR FIVE SIX  SEVENEIGHTNINE "
210   V$[46,65]="TEN  JACK QUEENKING "
220   S$="CLUBS   DIAMONDSHEARTS  SPADES  "
230   PRINT LIN(1)
235   REM Y IS SET 0 IF THE STAKE IS IN PRISON AND N IS SET 0
236   REM WHEN A COUP NEUTRE OCCURS.
240   Y=1
250   N=1
260   PRINT "HOW MUCH MONEY DO YOU HAVE ";
270   INPUT M
280   GOSUB 1490
290   PRINT TAB(13)"*** MESSIEURS, FAITES VOS JEUX...***"
300   PRINT LIN(1)
310   INPUT A$
320   PRINT LIN(1)
325   REM LOOP 330 TO 350 ENCODES THE INPUT IN 310 USING
326   REM THE TEST STRING G$ FROM 190.
330   FOR A=0 TO 7
340   IF A$[5,7]=G$[1+3*A,3+3*A] THEN 360
350   NEXT A
360   IF A>7 THEN 290
370   PRINT "HOW MUCH DO YOU WANT TO BET ";
380   INPUT B
390   IF B <= M THEN 420
400   PRINT "YOU DON'T HAVE ENOUGH MONEY TO COVER THAT BET !"
410   GOTO 370
420   PRINT LIN(1)
430   PRINT TAB(10)"*** LE JEU EST FAIT, RIEN NE VA PLUS...***"
440   PRINT LIN(1)
450   IF P<312 THEN 470
455   REM IN 450, 530, 680, THE COMPUTER CHECKS IF THERE ARE
456   REM ANY CARDS LEFT.
460   GOSUB 1470
465   REM C IS THE SUIT OF THE FIRST CARD IN THE UPPER ROW.
470   C=L[P+1]
480   PRINT "*** UPPER ROW ***"
485   REM S IS THE VALUE OF THE UPPER ROW, P IS THE NUMBER
486   REM OF CARDS ALREADY USED UP.
490   S=0
500   FOR J=1+P TO 312
510   S=S+G[J]
```

Fig. 1.5 Program "Trente-et-Quarante"

```
520    GOSUB 1850
530    IF J=312 THEN 460
540    IF S>30 THEN 560
550    NEXT J
560    K=J
570    IF S=40 THEN 600
580    PRINT "***    "S-30"   ***"
590    GOTO 610
600    PRINT "**   QUARANTE    **"
610    PRINT TAB(44)"*** LOWER ROW ***"
615    REM T IS THE VALUE OF THE LOWER ROW.
620    T=0
630    FOR J=K+1 TO 312
640    T=T+G[J]
650    PRINT TAB(44);
660    GOSUB 1850
670    IF T>30 THEN 700
680    IF J=312 THEN 460
690    NEXT J
700    F=J
710    IF T=S THEN 980
720    IF T=40 THEN 750
730    PRINT TAB(44)"***   "T-30"   ***"
740    GOTO 760
750    PRINT TAB(44)"**   QUARANTE    **"
755    REM IN 760 TO 1180, THE RESULT OF THE DEAL IS VERBA-
756    REM LIZED AND CODE NUMBERS ARE ASSIGNED TO THE OUTCOMES.
760    R$="ROUGE          ET COULEUR
770    IF T<S THEN 870
780    R$[7,10]="PERD"
790    E=0
800    IF C=1 OR C=2 THEN 840
810    R$[24,28]="GAGNE"
820    D=2
830    GOTO 950
840    D=3
850    R$[12,12]=","
860    GOTO 950
870    R$[7,11]="GAGNE"
880    E=1
890    IF C=1 OR C=2 THEN 930
900    R$[24,27]="PERD"
910    D=3
920    GOTO 950
930    R$[12,12]=","
940    D=2
950    PRINT LIN(1)
960    PRINT TAB(13)"*** "R$" ***"
970    GOTO 1210
980    IF T>31 THEN 1130
990    PRINT TAB(44)"*   UN   APRES   *"
1000   PRINT LIN(1)
1010   PRINT TAB(23)"*** REFAIT ***"
1020   PRINT LIN(1)
1030   IF A>3 OR Y=0 THEN 1290
1040   PRINT "DO YOU WANT HALF YOUR STAKE BACK ";
1050   INPUT Y$
1060   PRINT LIN(1)
1070   IF Y$[1,1]="N" THEN 1100
1080   B=B/2
```

Fig. 1.5 Program "Trente-et-Quarante" (cont'd)

```
1090    GOTO 1290
1100    Y=0
1110    PRINT TAB(12)"*** EN PRISON...HERE WE GO AGAIN ***"
1120    GOTO 440
1130    IF T=40 THEN 1160
1140    PRINT TAB(44)"*  "T-30" APRES *"
1150    GOTO 1170
1160    PRINT TAB(44)"*QUARANTE  APRES*"
1170    PRINT LIN(1)
1180    PRINT TAB(23)"*** COUP NUL ***"
1190    PRINT LIN(1)
1200    GOTO N*Y+1 OF 450,290
1210    PRINT LIN(1)
1215    REM IN 1220 TO 1280, THE COMPUTER CHECKS WHETHER THE
1216    REM OPERATOR LOST, BROKE EVEN, OR WON.
1220    IF Y=0 AND (A=E OR A=D) THEN 1390
1230    IF Y=0 THEN 1290
1240    IF A=E OR A=D OR A=2*D+E THEN 1360
1250    IF A=4 AND (E=0 OR D=2) THEN 1330
1260    IF A=5 AND (E=1 OR D=2) THEN 1330
1270    IF A=6 AND (E=0 OR D=3) THEN 1330
1280    IF A=7 AND (E=1 OR D=3) THEN 1330
1290    PRINT "YOU JUST LOST "B" LOUIS !"
1300    M=M-B
1310    IF M=0 THEN 1880
1320    GOTO 1400
1330    N=0
1340    PRINT TAB(12)"*** COUP NEUTRE...HERE WE GO AGAIN ***"
1350    GOTO 440
1360    PRINT "YOU JUST WON "B" LOUIS !"
1370    M=M+B
1380    GOTO 1400
1390    PRINT "YOU BROKE EVEN."
1400    PRINT "YOU NOW HAVE "M" LOUIS. WANT TO TRY IT AGAIN ";
1410    INPUT Y$
1420    PRINT LIN(1)
1430    IF Y$[1,1]="N" THEN 1910
1440    Y=1
1450    N=1
1460    GOTO 290
1470    PRINT LIN(1)
1480    PRINT TAB(19)"*** LES CARTES PASSENT ***"
1485    REM THE COMPUTER 'SHUFFLES' EACH OF 6 DECKS OF CARDS.
1486    REM S(J) IS THE NUMBER OF THE CARD IN POSITION J.
1490    PRINT LIN(1)
1500    PRINT TAB(7)"*** STAND BY -";
1510    PRINT " THE CARDS ARE BEING SHUFFLED ***"
1520    PRINT LIN(1)
1530    FOR K=0 TO 5
1540    FOR J=1+52*K TO 52+52*K
1550    S[J]=J
1560    NEXT J
1570    FOR I=52+52*K TO 2+52*K STEP -1
1580    J=INT(RND(1)*I)+1
1590    T=S[J]
1600    S[J]=S[I]
1610    S[I]=T
1620    NEXT I
1630    NEXT K
1635    REM THE COMPUTER 'SHUFFLES' THE 6 DECKS TOGETHER.
```

Fig. 1.5 Program "Trente-et-Quarante" (cont'd)

```
1640    FOR I=312 TO 2 STEP -1
1650    J=INT(RND(1)*I)+1
1660    T=S[J]
1670    S[J]=S[I]
1680    S[I]=T
1690    NEXT I
1700    PRINT TAB(1)"*** PLEASE CUT - ";
1710    PRINT "BY ENTERING A NUMBER BETWEEN 1 AND 312 ***"
1720    PRINT LIN(1)
1730    INPUT Z
1735    REM THE OPERATOR REMOVES THE TOP Z CARDS AND PUTS
1736    REM THEM ON THE BOTTOM.
1737    REM T(J) IS THE NUMBER OF THE CARD IN POSITION J
1738    REM AFTER THE CUT.
1740    FOR J=1 TO 312
1750    T[J]=S[J+Z-INT((J+Z-1)/312)*312]
1760    NEXT J
1770    FOR J=1 TO 312
1780    S[J]=T[J]-52*INT((T[J]-1)/52)
1790    V[J]=S[J]-13*INT((S[J]-1)/13)
1800    L[J]=INT((S[J]-1)/13)
1810    G[J]=V[J] MIN 10
1820    NEXT J
1830    P=0
1840    RETURN
1845    REM THE COMPUTER DEALS A CARD.
1850    PRINT V$[1+5*(V[J]-1),5+5*(V[J]-1)]" OF ";
1860    PRINT S$[1+8*L[J],8+8*L[J]]
1870    RETURN
1880    PRINT LIN(1)
1890    PRINT "YOU LOST ALL YOUR MONEY. BUZZ OFF !"
1900    GOTO 1930
1910    PRINT LIN(1)
1920    PRINT "GETTING COLD FEET ?"
1930    END
```

Fig. 1.5 Program "Trente-et-Quarante" (cont'd)

sub-routine; if it runs out of cards while dealing the lower row, the same result is accomplished by the instructions in line 680. Note that after reshuffling and cutting, the deal will resume with the first card of the first row, not where it left off. (Note also that the instructions in lines 670 and 680 are in reverse order from lines 530 and 540. Why?)

Y is the "prison variable" and N is the "coup neutre" variable. Normally 1, they are set to 0 when a stake is imprisoned or when a coup neutre occurs. This will cause the betting ballyhoo prior to the next deal to be bypassed. A zero value of Y also passes control to the "you lost . . . statement" in line 1290 in case of a refait when a stake is en prison.

# Chapter 2

# roulette

```
**********
*ROULETTE*
**********
```

DO YOU WANT TO PLAY AT THE EUROPEAN TABLE OR THE
AMERICAN TABLE ?EUROPEAN


HOW MUCH MONEY DO YOU HAVE ?10000
IN WHAT CURRENCY ?FRANCS


WHEN ASKED TO PLACE YOUR BETS ('MESSIEURS, FAITES VOS JEUX'),
ENTER ONE OF THE FOLLOWING:


```
        CHANCES SIMPLES (PAY-OFF:  1 TO 1)
        EN PLEIN        (PAY-OFF:35 TO 1)
        COLONNE         (PAY-OFF:  2 TO 1)
        DOUZAINE        (PAY-OFF:  2 TO 1)
        TRANSVERSALE    (PAY-OFF:11 TO 1)
        CARRE           (PAY-OFF:  8 TO 1)
        SIXAIN          (PAY-OFF:  5 TO 1)
        QUATRE PREMIERS (PAY-OFF:  8 TO 1)
```


```
             *** BONNE CHANCE ***
```



```
        *** HOW MUCH DO YOU WANT TO BET ? ***
```

?1000


```
        *** MESSIEURS, FAITES VOS JEUX ...***
```

Fig. 2.1 Run of the program "Roulette"

23

```
?CHANCES SIMPLES
A CHEVAL ? (ENTER YES OR NO) ?YES
ENTER 'ROUGE', OR 'NOIR', OR 'PAIR', OR 'IMPAIR',
OR 'MANQUE', OR 'PASSE' ?PAIR
AND ? (ENTER AN ADJACENT 'CHANCE SIMPLE')
?NOIR


          *** LES JEUX SONT FAITS, RIEN NE VA PLUS...***


             ***  31    ,NOIR  ,IMPAIR ET PASSE  ***


              *** COUP NEUTRE...HERE WE GO AGAIN ***


              ***  7      ,ROUGE ,IMPAIR ET MANQUE ***



YOU JUST LOST  1000     FRANCS !

YOU NOW HAVE  9000     FRANCS. DO YOU WANT TO TRY AGAIN ?YES


              *** HOW MUCH DO YOU WANT TO BET ? ***

?1000


              *** MESSIEURS, FAITES VOS JEUX ...***


?CHANCES SIMPLES
A CHEVAL ? (ENTER YES OR NO) ?YES
ENTER 'ROUGE', OR 'NOIR', OR 'PAIR', OR 'IMPAIR',
OR 'MANQUE', OR 'PASSE' ?MANQUE
AND ? (ENTER AN ADJACENT 'CHANCE SIMPLE')
?ROUGE
AND ? (ENTER AN ADJACENT 'CHANCE SIMPLE')
?IMPAIR


          *** LES JEUX SONT FAITS, RIEN NE VA PLUS...***


             ***  17    ,NOIR  ,IMPAIR ET MANQUE ***


     *** CONGRATULATIONS ! YOU JUST WON  1000     FRANCS !! ***


YOU NOW HAVE  10000   FRANCS. DO YOU WANT TO TRY AGAIN ?YES


              *** HOW MUCH DO YOU WANT TO BET ? ***

?2000
```

Fig. 2.1 Run of the program "Roulette" (cont'd)

```
        *** MESSIEURS, FAITES VOS JEUX ...***


?CHANCES SIMPLES
A CHEVAL ? (ENTER YES OR NO) ?NO
ENTER 'ROUGE', OR 'NOIR', OR 'PAIR', OR 'IMPAIR',
OR 'MANQUE', OR 'PASSE' ?NOIR


        *** LES JEUX SONT FAITS, RIEN NE VA PLUS...***


                *** ZERO ***

        *** DO YOU WANT HALF YOUR STAKE BACK ? ***


?NO
                *** EN PRISON...HERE WE GO AGAIN ***


        ***   1      ,ROUGE ,IMPAIR ET MANQUE ***



YOU JUST LOST  2000    FRANCS !

YOU NOW HAVE  8000     FRANCS. DO YOU WANT TO TRY AGAIN ?YES


        *** HOW MUCH DO YOU WANT TO BET ? ***
?100


                *** MESSIEURS, FAITES VOS JEUX ...***


?COLONNE
A CHEVAL ? (ENTER YES OR NO) ?YES
WHICH COLUMN ? (ENTER 1, 2, OR 3) ?2
AND WHICH ADJACENT COLUMN ?3


        *** LES JEUX SONT FAITS, RIEN NE VA PLUS...***


        ***  16     ,ROUGE ,PAIR   ET MANQUE ***



YOU JUST LOST  100   FRANCS !

YOU NOW HAVE  7900     FRANCS. DO YOU WANT TO TRY AGAIN ?YES


        *** HOW MUCH DO YOU WANT TO BET ? ***
?200
```

Fig. 2.1 Run of the program "Roulette" (cont'd)

```
              *** MESSIEURS, FAITES VOS JEUX ...***


?DOUZAINE
A CHEVAL ? (ENTER YES OR NO) ?YES
WHICH DOZEN ?
ENTER 'PREMIERE', MOYENNE', OR 'DERNIERE' ?MOYENNE
AND WHICH ADJACENT DOZEN ?DERNIERE


         *** LES JEUX SONT FAITS, RIEN NE VA PLUS...***


           ***   21    ,ROUGE ,IMPAIR ET PASSE   ***


     *** CONGRATULATIONS ! YOU JUST WON   100   FRANCS !! ***


YOU NOW HAVE  8000    FRANCS. DO YOU WANT TO TRY AGAIN ?YES


              *** HOW MUCH DO YOU WANT TO BET ? ***
?100


              *** MESSIEURS, FAITES VOS JEUX ...***


?TRANSVERSALE
WHICH ROW DO YOU WANT TO BACK ?
ENTER 0, OR 1, OR 2, ..., OR 12 ?0
WITH 1-2, OR WITH 2-3 ?
ENTER THE TWO NUMBERS, ONE AT A TIME ?2
?3


         *** LES JEUX SONT FAITS, RIEN NE VA PLUS...***


           ***   3    ,ROUGE ,IMPAIR ET MANQUE ***


     *** CONGRATULATIONS ! YOU JUST WON   1100     FRANCS !! ***


YOU NOW HAVE  9100    FRANCS. DO YOU WANT TO TRY AGAIN ?YES


              *** HOW MUCH DO YOU WANT TO BET ? ***
?1000


              *** MESSIEURS, FAITES VOS JEUX ...***


?CARRE
WHICH SQUARE ?
ENTER THE NUMBER FROM THE LEFT UPPER CORNER ?19
```

Fig. 2.1 Run of the program "Roulette" (cont'd)

```
        *** LES JEUX SONT FAITS, RIEN NE VA PLUS...***


            ***   30     ,ROUGE ,PAIR    ET PASSE   ***



YOU JUST LOST   1000      FRANCS !


YOU NOW HAVE   8100      FRANCS. DO YOU WANT TO TRY AGAIN ?YES


              *** HOW MUCH DO YOU WANT TO BET ? ***
?1000


              *** MESSIEURS, FAITES VOS JEUX ...***


?SIXAIN
WHICH IS THE FIRST OF THE TWO ADJACENT ROWS ?
ENTER 1, OR 2, OR 3,..., OR 11 ?4


            *** LES JEUX SONT FAITS, RIEN NE VA PLUS...***


            ***   34     ,ROUGE ,PAIR    ET PASSE   ***

YOU JUST LOST   1000      FRANCS !


YOU NOW HAVE   7100      FRANCS. DO YOU WANT TO TRY AGAIN ?YES


              *** HOW MUCH DO YOU WANT TO BET ? ***
?1000


              *** MESSIEURS, FAITES VOS JEUX ...***


?QUATRE PREMIERS


            *** LES JEUX SONT FAITS, RIEN NE VA PLUS...***


            ***   12     ,ROUGE ,PAIR    ET MANQUE ***



YOU JUST LOST   1000      FRANCS !


YOU NOW HAVE   6100      FRANCS. DO YOU WANT TO TRY AGAIN ?YES
```

*Fig. 2.1 Run of the program "Roulette" (cont'd)*

```
        *** HOW MUCH DO YOU WANT TO BET ? ***

?1000


        *** MESSIEURS, FAITES VOS JEUX ...***

?EN PLEIN
A CHEVAL ? (ENTER YES OR NO) ?NO
ENTER ONE OF THE NUMBERS 0,1,2,3,...,36 ?11


     *** LES JEUX SONT FAITS, RIEN NE VA PLUS...***


       ***  1     ,ROUGE ,IMPAIR ET MANQUE ***



YOU JUST LOST  1000     FRANCS !

YOU NOW HAVE  5100     FRANCS. DO YOU WANT TO TRY AGAIN ?YES


        *** HOW MUCH DO YOU WANT TO BET ? ***

?5100


        *** MESSIEURS, FAITES VOS JEUX ...***

?EN PLEIN
A CHEVAL ? (ENTER YES OR NO) ?YES
ENTER ONE OF THE NUMBERS 0,1,2,3,...,36 ?19
AND AN ADJACENT NUMBER ?22


     *** LES JEUX SONT FAITS, RIEN NE VA PLUS...***

       *** 10    ,NOIR  ,PAIR   ET MANQUE ***



YOU JUST LOST  5100     FRANCS !

MOTIVATED BY GREED AND CURSED WITH CONGENITAL STUPIDITY, YOU
JUST LOST ALL YOUR MONEY. YOU EITHER TAKE THE CONSEQUENCES
LIKE A GENTLEMEN, OR YOU WILL HAVE TO WORK OFF YOUR BAR-BILL
IN THE KITCHEN !

DONE
```

*Fig. 2.1 Run of the program "Roulette" (cont'd)*

When we produced the preceding printout (Fig. 2.1), it was not our objective to "make a killing" but rather to demonstrate the versatility and adaptability of our computer program.

The reader who already knows how to play roulette may proceed directly to page 40. The others will have to catch up by working their way through the following sections.

*"You cannot beat the roulette wheel without stealing money from the table."*
ALBERT EINSTEIN

## HOW THE GAME IS PLAYED

Roulette is played at a roulette table that is associated with a roulette wheel *(le cylindre)*. In Fig. 2.2(a), we display the layout of the Monte Carlo (European) roulette table, and in Fig. 2.2(b), the layout of the Las Vegas-Atlantic City (American) roulette table. In European roulette there is, in fact, another table with exactly the same layout on the other side of the wheel. Two *croupiers* (housemen) sit on each side of the wheel, taking care of the betting on the side closest to them. At the American table, there is usually no such second wing. The wheel and the right side of the table are roped off, and two attendants, positioned in the roped-off section, attend to the necessary chores.

At the European table, a *croupier* (attendant) invites the players *(les suckers)*, who sit, lounge, or stand around the roulette table, to place their bets, intoning

★★★ *Messieurs, faites vos jeux* ★★★

(Gentlemen, place your bets). He then spins the wheel and throws a small ivory ball at the sloping side of the basin in a direction opposite to the spin of the wheel. The players place their bets by putting chips (slugs) onto the appropriate field on the roulette table. To bet on *rouge* (red), for example, one places the chip(s) on the field marked by the red diamond; to bet on 19, one places the chip(s) on 19, and so forth. As the ivory ball loses its momentum and starts bouncing around erratically, the *croupier* puts an end to the betting with

★★★ *Les jeux sont faits, rien ne va plus* ★★★

(The betting is over, no more bets). Now, everybody waits anxiously for the ball to come to rest. As soon as that happens, the *croupier* announces the result by first stating the winning number, then the

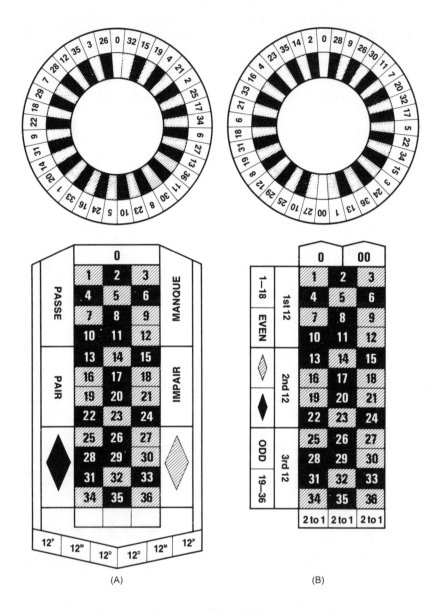

*Fig. 2.2 (a) European table, and (b) American table*

color, then whether the number is even or odd, and, finally, whether it is low or high. For example, if the ball comes to rest on 29, the croupier chants in a monotonous voice

★★★ *Vingt-neuf, noir, impair, et passe* ★★★

(twenty-nine, black, odd, and high). The loser's stakes are raked in, the winners are paid (one hopes), and the next game commences. The whole process lasts about two minutes at the European table, one-half minute at the American table. The faster pace at the American table is achieved by dispensing with the ceremonial incantations in French. French is used, however, at most European casinos, no matter what country they operate in.

For those who intend to try their luck in Europe or South Africa, we will explain the various terms as we move along. The European wheel is subdivided into 37 compartments of identical size, these compartments being numbered from 0 to 36 [see Fig. 2.3(a)], whereas the American wheel is subdivided into 38 such compartments with an extra field, labeled 00, thrown in [see Fig. 2.3(b)]. Henceforth, we shall

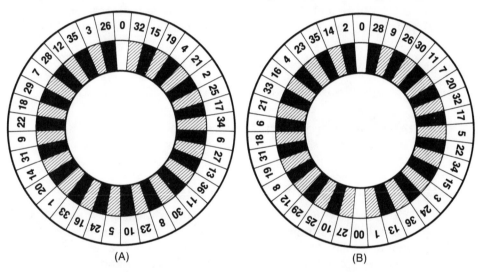

(A)                                        (B)

Fig. 2.3 (a) European wheel, and (b) American wheel

use the term *European table* in the following strict sense: a layout as shown in Fig. 2.2(a) and a wheel with 37 compartments as shown in Fig. 2.3(a). *American table* shall mean a layout as shown in Fig. 2.2(b) and a wheel with 38 compartments as shown in Fig. 2.3(b). The English table and the South African table have the same layout as the American Table except that the fields for 00 and 0 are combined into one field labeled 0 (see Fig. 2.4) and their wheel is the same as that for the European table. Hence, for all practical purposes, they are equivalent to the European table and will be treated as such.

The numbers on the European wheel alternate between *manque* (low, namely the numbers 1, 2, 3, . . . , 18) and *passe* (high, namely the numbers 19, 20, 21, . . . , 36). On both wheels, *rouge* (red) alternates

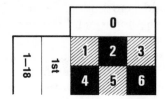

Fig. 2.4 *Layout of English and South African tables*

with *noir* (black). Although the numbers are arranged differently on the two wheels, red and black are assigned to the same numbers on both. The zero and the double-zero are white, green, or blue.

Every slot (compartment) is thus identified by a number that is either 0 (or 00) or not. If not, then it is either *noir* (black) or *rouge* (red), *pair* (even) or *impair* (odd), and *passe* (high) or *manque* (low).

## WHY THE GAME IS PLAYED

You can't take it with you, *n'est ce pas?*

## *"Les mises sur paroles sont rigoureuse-ment interdites"*[1]

## WHAT TO BET AND HOW TO GO ABOUT IT

One may bet on *noir, rouge, pair, impair, passe, manque,* a specific number, certain groups of numbers, or some combinations of the aforementioned. Not every gaming establishment allows all the bets which we are going to discuss. But, as far as we know, there are no bets allowed anywhere—with the possible exception of some isolated villages on the upper Mekong—which are not described here. The pay-off and the minimum and maximum bets allowed depend on the odds and on the house. The pay-off always favors the house, at the American table to a much greater extent than at the European table. Norman Squire, one of the leading gambling experts of our times, says that no serious player will play roulette at an American table (Ref. [12], p. 25). We drink to that! Bob Martin, gambling authority and official sports odds-maker from Las Vegas says on the same subject: "You've gotta get lucky fast and quit fast, while you're still ahead. If you ever see me at a roulette table, you'll know I'm playing for kicks . . ." (Ref. [7], p. 106).

---

[1] Announced bets not accompanied by chips are strictly prohibited.

We are now ready to discuss all possible bets. Figure 2.5 illustrates how to place each bet physically.

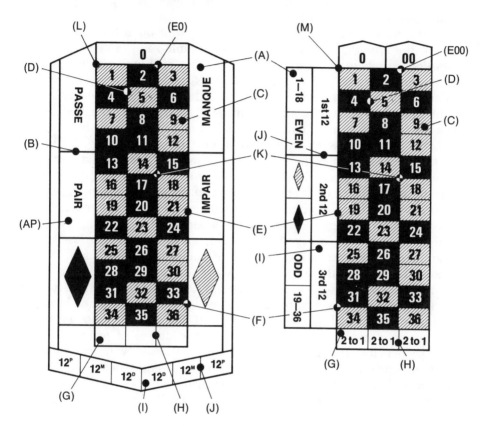

*Fig. 2.5 Placement of chips*

### Chances Simples (Even Chances)

*Pay-off:*     Even money
*Odds for winning:*     18:19 at the European table, 18:20 at the American table
*Expected loss:*     $13.51 per $1000 bet at the European table
$52.63 per $1000 bet at the American table ($26.31 if half the stake is returned on 0 and 00)

(Actually, *chances simples* is a misnomer. The chances are, as seen from the above data, somewhat less than even!)

You may put your stake (*la mise*) on *noir* (black), *rouge* (red), *pair* (even), *impair* (odd), *passe* (high), or *manque* (low) by placing your chip(s) on the field so marked. [See (A) in Fig. 2.5 for the case of a bet on *manque*, that is, one of the numbers 1, 2, 3, . . . , 18.]

If you win the *coup* (turn), you are paid even money; if you lose, your stake is raked in by the *croupier*.

If the ball comes to rest on 0 (or either 0 or 00 at the American table), the *croupier* announces *zéro* (or double-zero), and then, depending on where you are playing, a number of things may happen. At the American table, the croupier will, as a rule, rake in your stake. There is, however, at least one casino in Atlantic City that returns half your stake on 0 and 00 for all 1–1 "ouside" bets. [A 1–1 "outside" bet is a bet for even money that is placed on one of the outside fields on the table, that is, low, even, red, black, odd, or high. See Fig. 2.2(b).] At the European table, the croupier may rake in half your stake, he may give you a choice of his doing this or of putting your money *en prison* (in prison), or he may simply put your money *en prison* without giving you a choice. If your bet is placed *en prison*, then your chips are placed on the line bordering the outside of the field they were sitting on [see (AP) in Fig. 2.5 for a *pair en prison*]. If you lose the next *coup*, your stake is forfeited, but if you win the next *coup*, it is liberated, that is, put back where it was. You may let it ride for the next *coup*, you may place it elsewhere, or, for that matter, you may pocket it and walk over to the baccarat table to ogle the blonde broad with the low-cut black evening gown. If the zero is followed by another zero and the house allows only one degree of prison, then your stake is lost. If two degrees of prison are allowed, your stake goes to double-prison. If you win the next turn, it goes back to simple prison, and you proceed as above. Otherwise, your stake is lost. Some casinos allow three degrees of prison. Big deal! How often do you think that three zeros will appear in a row?

In another variation of the prison option, your stake stays in prison until a result other than zero obtains. When that happens, you either get your stake back or you lose your stake.

If you have a choice between getting half your stake back and having your stake imprisoned (no matter how may degrees of prison are offered), by all means take back half your stake. You can't do any better (see also Appendix A, section 7, and Ref. [9]).

In our computer simulation, we give the player a choice of either forfeiting half his stake or going to prison, and we allow prison of only one degree.

Since you will break even, at best, if a zero comes up, your chances of winning are somewhat less than even. In fact, the *expected value of the game* at the European table is

$$E_e = -N/74 = -0.0135135\ N$$

This means that in the long run (whatever that means) you will lose
N/74 dollars (or francs, pounds, yen, or whatever the currency may be)
for a total investment of N dollars (or whatever). Note that, due to the
presence of the 00 at the customary American table and the automatic
loss in case of a 0 or 00, the expected value at the American table is

$$E_a = -2N/38 = -0.052631\,N$$

or almost four times as much!

In the short run, of course, you may win or lose several million!
If you have a run of luck and win a lot of money, you are not honor-
bound to keep on playing and give the casino a chance to win it all
back. Don't be stupid! Blow it on booze and broads, pay up the old
mortgage, or do something rash. Only remember that if you are a U.S.
citizen or resident, you will have to share it with Uncle Sam. Be
careful. For all you know, the blonde dame with the low-cut black
evening gown at the baccarat table might be an IRS undercover agent.
Of course, you could consider selling the split-level in the New Jersey
swamps and taking up permanent residence in Switzerland or some-
place (Albania is not recommended; it's worse than New Jersey!).

### À cheval Bets on Chances Simples (Split Bets on Even Chances)

*Pay-off:*     Even money
*Odds for winning:*     10:11 on *rouge-impair* vs. *pair-noir*
                        9:10 on *impair-manque* vs. *passe-pair*
*Expected loss:*     $27.03 per $1000 bet at the European table

This bet is usually not accepted at the American table and the
English table. (The odds given are those at the European Table for
winning vs. losing. The odds for winning vs. not winning would be
10:27 and 9:28, respectively.)

You may bet *à cheval* (astride) on any of these two adjacent
*chances simples*:

| | |
|---|---|
| *passe–pair* | *rouge–impair* |
| *pair–noir* | *impair–manque* |

by placing your chip(s) on the line separating the two fields [that's
why it is called *à cheval*; see (B) in Fig. 2.5 for a bet on *passe–pair*].

If *zéro* comes up, your bet is forfeited. If both win, you win even
money. If both lose, you lose. In case of a *coup neutre* (one wins and
the other loses), you break even and may withdraw your stake or leave
it in place for the next turn.

Since the chances for *passe-pair* vs. *impair-manque* and for *pair-
noir* vs. *rouge-impair* are exactly even, only the probability for the

occurrence of a zéro enters the calculations of the *expected value of the game*, which is thus

$$E = -\ N/37 = -\ 0.027027\ N$$

Why bet *à cheval* if the expected loss is twice as much as on *chances simples*? "Because it's there" is just as good an answer as any.

### En Plein (Straight Up)

Pay off:     35:1
Odds for winning:     1:36 at the European table, 1:37 at the American table
Expected loss:     $27.03 per $1000 bet at the European table
$52.63 per $1000 bet at the American table

You may bet on any number including 0 (and 00 at the American table) by placing your chip(s) on the appropriate field [see (C) in Fig. 2.5 for a bet on the number 9].

The *expected value* of this game is

$$E_e = -\ N/37 = -\ 0.027027\ N\ \text{at the European table}$$

$$E_a = -\ 2N/38 = -\ 0.052631\ N\ \text{at the American table}$$

What motivates gamblers to bet *en plein*? Greed!

### À Cheval (Split Bet)

Pay-off:     17:1
Odds for winning:     2:35 at the European table, 2:36 at the American table
Expected loss:     $27.03 per $1000 bet at the European table
$52.63 per $1000 bet at the American table

You may bet on any two adjacent numbers (including 0–1, 0–2, 0–3 at the European table and 0–00, 0–1, 0–2, 00–2, and 00–3 at the American table) by placing your chip(s) on the dividing line between the two numbers [see (D) in Fig. 2.5 for an *à cheval* bet on 4–5].

The expected value of this game and all following ones, except *cinq premiers*, is the same as for *en plein*.

What motivates gamblers to bet *à cheval*? Greed, tempered by caution!

### Transversale (Row, Down the Street, Down the Side)

Pay-off:     11:1
Odds for winning:     3:34 at the European table, 3:35 at the American table

*Expected loss:*     $27.03 per $1000 bet at the European table
                    $52.63 per $1000 bet at the American table

You may bet on any of the rows, such as 1–2–3, 4–5–6, etc., by placing your chip(s) on the line dividing the row from the adjacent *chance simple* field at the European table and from the adjacent *dozen* field at the American table [see (E) in Fig. 2.5 for a bet on the row 19–20–21]. To bet on "rows" 0–1–2 or 0–2–3 at the European table, you place your chip(s) as indicated by (E0) in Fig. 2.5 for a bet on "row" 0–2–3. To bet on "rows" 0–1–2, 00–0–2, or 00–2–3 at the American table, you place your chip(s) as indicated by (E00)) in Fig. 2.5 for a bet on "row" 00–2–3. (Incidentally, the last row, 34–35–36, is often referred to as *les trois dernières,* or "the last three.")

## Sixain (Sixline, an À Cheval Bet on Two Adjacent Transversales)

*Pay-off:*     5:1
*Odds for winning:*     6:31 at the European table, 6:32 at the American table
*Expected loss:*     $27.03 per $1000 bet at the European table
                    $52.63 per $1000 bet at the American table

You may bet on two adjacent rows by placing your chip(s) on the line dividing the two rows, as indicated by (F) in Fig. 2.5 for a bet on the two rows 31–32–33 and 34–35–36. (At the American table, the chip has to be placed on the "dozen side." At the European table, the side does not matter.) The first two rows, 1–2–3 and 4–5–6, are called *les six premiers,* or "the first six."

## Colonne (Column)

*Pay-off:*     2:1
*Odds for winning:*     12:25 at the European table, 12:26 at the American table
*Expected loss:*     $27.03 per $1000 bet at the European table
                    $52.63 per $1000 bet at the American table

You may bet on a column by placing you chip(s) as indicated in Fig. 2.5 by (G) for a bet on column number 1. You win if one of the numbers in your column comes up.

## À Cheval on Two Adjacent Colonnes (Split Bet on Two Adjacent Columns)

*Pay-off:*     1:2
*Odds for winning:*     24:13 at the European table, 24:14 at the American table

*Expected loss:*      $27.03 per $1000 bet at the European table
$52.63 per $1000 bet at the American table

You may bet on two adjacent columns by placing your chip(s) as indicated in Fig. 2.5 by (H) for an *à cheval* bet on the second and third columns.

### Douzaine (Dozen)

*Pay-off, Odds, Expected loss:*      Same as for *colonne*

You may bet on *la première douzaine* (the first dozen), *la moyenne (deuxième) douzaine* (the middle or second dozen), or *la dernière douzaine* (the last dozen) by placing your chip(s) as indicated in Fig. 2.5 by (I) for a bet on *la dernière douzaine*.

### À Cheval on Two Adjacent Douzaines (Split Bet on Two Adjacent Dozens)

*Pay-off, Odds, Expected loss:*      Same as for two adjacent
*colonnes*

You may bet *à cheval* on two adjacent dozens by placing your chip(s) as indicated in Fig. 2.5 by (J) for an *à cheval* bet on *la première douzaine* and *la moyenne douzaine*.

### Carré (Corner Bet)

*Pay-off:*      8:1
*Odds for winning:*      4:33 at the European table, 4:34 at the American table
*Expected loss:*      $27.03 per $1000 bet at the European table
$52.63 per $1000 bet at the American table

You may bet on a group of four numbers that form a square, such as 14–15–17–18 by placing your chip(s) as indicated in Fig. 2.5 by (K).

### Quatre Premiers (First Four)

*Pay-off, Odds, Expected loss:*      Same as for *carré*

You may bet on the first four numbers, 0–1–2–3, by placing your chip(s) as indicated in Fig. 2.5 by (L).

This bet as such is not accepted at the American table. You may, of course, place two *à cheval* bets, one on 0–1, the other on 2–3.

**Cinq Premiers (First Five)**

*Pay-off:*     6:1
*Odds for winning:*     5:33 at the American table
*Expected loss:*     $78.94 per $1000 bet at the American table

You may bet on the first five numbers, 0–00–1–2–3, by placing your chip(s) as indicated in Fig. 2.5 by (M).

This bet is not possible at the European table, which is no great loss because it is a sucker bet by any standards.

Of course, you may also place any combination of the aforementioned bets so long as they are accepted at your casino. Some popular combinations have special names, for example, *le tiers du cylindre* (one third of the wheel), which consists of the numbers 27, 13, 36, 11, 30, 8, 23, 10, 5, 24, 16, 33 (see also Fig. 2.6). In order to make such a bet, you need six chips (or a multiple of six chips), which are placed in equal amounts *à cheval* on the combinations 5–8, 10–11, 13–16, 23–24, 27–30, 33–36. The payoff, of course, is 2:1.

Another special combination is *les voisins du zéro* (the neighbors of zero), or the numbers 22, 18, 29, 7, 28, 12, 35, 3, 26, 0, 32, 15, 19, 4, 21, 2, 25 (see also Fig. 2.6). Nine chips (or a multiple of nine chips)

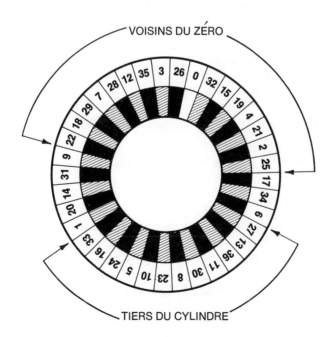

*Fig. 2.6 Combination bets: voisins du zéro and tiers du cylindre*

are required to place a bet on these numbers. Two parts of your stake are placed on the row 0–2–3 [as explained under *transversale* and indicated in Fig. 2.5 by (E0)] and the other seven parts are placed *à cheval* on the combinations 4–7, 12–15, 18–21, 19–22, 25–28, 26–29, and 32–35. The odds for winning are 17:19.

Finally, let us mention *les orphelins* (the orphans), which are the numbers that are left over after *le tiers* and *les voisins du zéro* have been taken out, namely, 1, 20, 14, 31, 9, 17, 34, and 6. The odds for winning are 8:29.

With these explanations out of the way, we are ready to try some simple systems or join the blonde broad with the low-cut black evening gown at the baccarat table—but *not* both. Gambling is serious business.

*"A bad system is better than none at all."*
SIR HIRAM MAXIM (1840–1916)[1]

## SOME SIMPLE SYSTEMS

The data provided in the preceding section demonstrate beyond any reasonable doubt that the odds are against you and that you are bound to lose—eventually. However, if you are sober, restrained, not greedy, very lucky, if money does not mean all that much to you anyway, and if you can forget the blonde broad with the low-cut black evening gown at the baccarat table for the time being, you may be able to win a little by playing judiciously.

*"You have not played yet? Do not do so; above all avoid a martingale if you do."*
WILLIAM MAKEPEACE THACKERAY (1811–1863)[2]

### The Martingale

The *martingale* is the most widely known system for betting on a *chance simple* (even chance). It has the reputation of being the ticket to certain success! How could you possibly lose if you double your bet after every loss? You bet, say, 20 francs and lose. Big deal, so you are 20 francs in the hole. You merely double your bet to 40 francs, and if you win at that turn, you are 20 francs ahead. If you lose, you are 60

[1] Inventor of the first automatic repeating gun.
[2] English novelist.

francs in the hole, but you double your bet to 80 francs, and, if you win, you are 20 francs ahead. If you lose, you are 140 francs in the hole, but you double your bet to 160 francs, which, as before, gives you a chance to recoup your losses and wind up 20 francs ahead, and so on. Perfect, no? No! To be sure, the system would be perfect if

1. You had an unlimited supply of capital
2. You had all the time in the world
3. There were no limit to what you are allowed to bet.

Note that all three conditions must be met to make the martingale what it is reputed to be. They never are. Even though you may have an unlimited amount of money and, concomitantly, all the time in the world, there always is a *house maximum*, the largest amount you may bet at any one time on any one chance. For example, in Monte Carlo, the house maximum on a *chance simple* at trente-et-quarante is 20,000 francs (1000 louis) in the "kitchen" and 50,000 francs (2500 louis) in the *salles privées* (private rooms), while the *house minimum* (the smallest admissible bet) is 20 francs (1 louis) in the "kitchen" and 50 francs (2½ louis) in the *salles privées*. For roulette, the house minimum and house maximum for bets on even chances are 10 francs and 10,000 francs, respectively, in the kitchen, and 20 francs and 20,000 francs, respectively, in the *salles privées*. (These figures may have been raised since 1977.) In Atlantic City, the minimum is $2 at some tables and $5 at others. The house maximum is $2,000 on any 1–1 bet.

Table 2.1 illustrates what can happen in the Monte Carlo *salles privées* when one starts out with a 20-franc bet on *noir* and plays a martingale progression. After *coup* No. 10, you were already 20,460 francs in the hole, you had to cough up another 20,000 francs for the next bet, you were lucky and won but are still 460 francs in the hole! A perfect system? Horsefeathers!

**Table 2.1  Example of a martingale progression**

| Coup No. | Bet | Result | Win (Lose) | Net gain (Loss) |
|---|---|---|---|---|
| 1 | 20 | 9 | (20) | (20) |
| 2 | 40 | 27 | (40) | (60) |
| 3 | 80 | 34 | (80) | (140) |
| 4 | 160 | 7 | (160) | (300) |
| 5 | 320 | 18 | (320) | (620) |
| 6 | 640 | 14 | (640) | (1260) |
| 7 | 1280 | 32 | (1280) | (2540) |
| 8 | 2560 | 21 | (2560) | (5100) |
| 9 | 5120 | 5 | (5120) | (10220) |
| 10 | 10240 | 16 | (10240) | (20460) |
| 11 | 20000 | 13 | 20000 | (460) |

Suppose that instead of winning *coup* No. 11, you don't win until *coup* No. 12, as shown below:

| 10 | 10240 | 12 | (10240) | (20460) |
|----|-------|----|---------|---------|
| 11 | 20000 | 19 | (20000) | (40460) |
| 12 | 20000 | 31 | 20000   | (20460) |

You needed a capital of 60,460 francs only to end up 20,460 francs (about $5,000) in the hole! Don't you think that Thackeray had a point?

We went on the computer and played a martingale progression on red, starting with a two dollar bet. The results are shown in Fig. 2.7.

```
        *********
        *ROULETTE*
        *********
```

```
DO YOU WANT TO PLAY AT THE EUROPEAN TABLE OR THE
AMERICAN TABLE ?AMERICAN

HOW MUCH MONEY DO YOU HAVE ?2000

WHEN ASKED TO PLACE YOUR BETS ('MESSIEURS, FAITES VOS JEUX'),
ENTER ONE OF THE FOLLOWING:

            EVEN CHANCE     (PAY-OFF: 1 TO 1)
            STRAIGHT UP     (PAY-OFF:35 TO 1)
            COLUMN          (PAY-OFF: 2 TO 1)
            DOZEN           (PAY-OFF: 2 TO 1)
            STREET          (PAY-OFF:11 TO 1)
            CORNER          (PAY-OFF: 8 TO 1)
            SIXLINE         (PAY-OFF: 5 TO 1)
            FIRST FIVE      (PAY-OFF: 6 TO 1)


            *** GOOD LUCK ***



    *** HOW MUCH DO YOU WANT TO BET ? ***

?2


    *** MESSIEURS, FAITES VOS JEUX ...***
```

Fig. 2.7 Run of a martingale progression on red

```
?EVEN CHANCE
ENTER 'RED', OR 'BLACK', OR 'EVEN', OR 'ODD',
OR 'LOW', OR 'HIGH' ?RED


          *** LES JEUX SONT FAITS, RIEN NE VA PLUS...***


            ***  11    ,BLACK ,ODD    AND LOW    ***



YOU JUST LOST  2     DOLLARS !

YOU NOW HAVE  1998    DOLLARS. DO YOU WANT TO TRY AGAIN ?YES


            *** HOW MUCH DO YOU WANT TO BET ? ***
?4


            *** MESSIEURS, FAITES VOS JEUX ...***


?EVEN CHANCE
ENTER 'RED', OR 'BLACK', OR 'EVEN', OR 'ODD',
OR 'LOW', OR 'HIGH' ?RED


          *** LES JEUX SONT FAITS, RIEN NE VA PLUS...***


            ***  29    ,BLACK ,ODD    AND HIGH   ***



YOU JUST LOST  4     DOLLARS !

YOU NOW HAVE  1994    DOLLARS. DO YOU WANT TO TRY AGAIN ?YES


            *** HOW MUCH DO YOU WANT TO BET ? ***
?8


            *** MESSIEURS, FAITES VOS JEUX ...***


?EVEN CHANCE
ENTER 'RED', OR 'BLACK', OR 'EVEN', OR 'ODD',
OR 'LOW', OR 'HIGH' ?RED


          *** LES JEUX SONT FAITS, RIEN NE VA PLUS...***
```

*Fig. 2.7 Run of a martingale progression on red (cont'd)*

```
          ***  31   ,BLACK ,ODD    AND HIGH   ***



YOU JUST LOST  8     DOLLARS !

YOU NOW HAVE  1986     DOLLARS. DO YOU WANT TO TRY AGAIN ?YES

          *** HOW MUCH DO YOU WANT TO BET ? ***
?16


          *** MESSIEURS, FAITES VOS JEUX ...***


?EVEN CHANUCE
ENTER 'RED', OR 'BLACK', OR 'EVEN', OR 'ODD',
OR 'LOW', OR 'HIGH' ?RED


       *** LES JEUX SONT FAITS, RIEN NE VA PLUS...***

          *** 2    ,BLACK ,EVEN   AND LOW    ***



YOU JUST LOST  16    DOLLARS !


YOU NOW HAVE  1970     DOLLARS. DO YOU WANT TO TRY AGAIN ?YES

          *** HOW MUCH DO YOU WANT TO BET ? ***
?32


          *** MESSIEURS, FAITES VOS JEUX ...***


?EVEN CHANCE
ENTER 'RED', OR 'BLACK', OR 'EVEN', OR 'ODD',
OR 'LOW', OR 'HIGH' ?RED


       *** LES JEUX SONT FAITS, RIEN NE VA PLUS...***

          *** 26   ,BLACK ,EVEN   AND HIGH   ***



YOU JUST LOST  32    DOLLARS !
```

Fig. 2.7 Run of a martingale progression on red (cont'd)

YOU NOW HAVE  1938     DOLLARS. DO YOU WANT TO TRY AGAIN ?YES

                    *** HOW MUCH DO YOU WANT TO BET ? ***

?64

                    *** MESSIEURS, FAITES VOS JEUX ...***

?EVEN CHANCE
ENTER 'RED', OR 'BLACK', OR 'EVEN', OR 'ODD',
OR 'LOW', OR 'HIGH' ?RED

          *** LES JEUX SONT FAITS, RIEN NE VA PLUS...***

              ***  8      ,BLACK ,EVEN    AND LOW    ***


YOU JUST LOST  64    DOLLARS !

YOU NOW HAVE  1874     DOLLARS. DO YOU WANT TO TRY AGAIN ?YES

                    *** HOW MUCH DO YOU WANT TO BET ? ***

?128

                    *** MESSIEURS, FAITES VOS JEUX ...***

?EVEN CHANCE
ENTER 'RED', OR 'BLACK', OR 'EVEN', OR 'ODD',
OR 'LOW', OR 'HIGH' ?RED

          *** LES JEUX SONT FAITS, RIEN NE VA PLUS...***

              ***  4      ,BLACK ,EVEN    AND LOW    ***

YOU JUST LOST  128   DOLLARS !

YOU NOW HAVE  1746     DOLLARS. DO YOU WANT TO TRY AGAIN ?YES

                    *** HOW MUCH DO YOU WANT TO BET ? ***

?256

                  *** MESSIEURS, FAITES VOS JEUX ...***

Fig. 2.7 Run of a martingale progression on red (cont'd)

```
?EVEN CHANCE
ENTER 'RED', OR 'BLACK', OR 'EVEN', OR 'ODD',
OR 'LOW', OR 'HIGH' ?RED


        *** LES JEUX SONT FAITS, RIEN NE VA PLUS...***


            ***  28   ,BLACK ,EVEN   AND HIGH   ***



YOU JUST LOST  256   DOLLARS !

YOU NOW HAVE  1490    DOLLARS. DO YOU WANT TO TRY AGAIN ?YES

            *** HOW MUCH DO YOU WANT TO BET ? ***
?512


            *** MESSIEURS, FAITES VOS JEUX ...***


?EVEN CHANCE
ENTER 'RED', OR 'BLACK', OR 'EVEN', OR 'ODD',
OR 'LOW', OR 'HIGH' ?RED


        *** LES JEUX SONT FAITS, RIEN NE VA PLUS...***


            ***  36   ,RED   ,EVEN   AND HIGH   ***


        *** CONGRATULATIONS ! YOU JUST WON   512   DOLLARS !! ***

YOU NOW HAVE  2002    DOLLARS. DO YOU WANT TO TRY AGAIN ?NO

GETTING COLD FEET ? AFTER ALL, IT'S ONLY MONEY !

DONE
```

Fig. 2.7 Run of a martingale progression on red (cont'd)

## Grande Martingale

Since the martingale is not the panacea it was first believed to be, all kinds of modifications have been proposed. There is, for example, the grande martingale, in which you not only double your bet after each loss but also augment it by your original basic bet, betting in succession 20, 60, 140, 300, 620, 1,260, 2,540, 5,100, and

10,220 francs. In this manner, you will reach the house maximum one *coup* earlier than before and hence run a greater risk of losing a lot of money. However, if you win any time after the first *coup*, your reward is more than a measly 20 francs. At the second *coup*, it is 40 francs, at the third *coup* 60, at the fourth 80, then 100, 120, 140, 160, and, finally, 180 at the ninth *coup*. If you lose the ninth (valid) *coup* and all the *coups* before it, you are 20,260 francs in the hole and cannot recoup with one bet because of the house maximum.

## Boule de Neige

The *boule de neige* (snowball) is a *martingale in reverse*. You double your stake only after each win and quit when you have lost as much as you can afford or won enough to satisfy your greed. Suppose you can afford to lose 100 francs. Table 2.2 shows what might happen when you bet on *manque* (low).

Table 2.2   Example of boule de neige system

| Coup No. | Bet | Result | Win (Lose) | Net gain (Loss) |
|---|---|---|---|---|
| 1 | 20 | 21 | (20) | (20) |
| 2 | 20 | 33 | (20) | (40) |
| 3 | 20 | 27 | (20) | (60) |
| 4 | 20 | 3 | 20 | (40) |
| 5 | 40 | 17 | 40 | 0 |
| 6 | 80 | 6 | 80 | 80 |

We tried the *boule de neige* system on the computer, backing *manque*, starting out with 20 francs, and having 100 francs to play with. The results are shown in Fig. 2.8.

```
DO YOU WANT TO PLAY AT THE EUROPEAN TABLE OR THE
AMERICAN TABLE ?EUROPEAN


HOW MUCH MONEY DO YOU HAVE ?100
IN WHAT CURRENCY ?FRANCS



                *** BONNE CHANCE ***



        *** HOW MUCH DO YOU WANT TO BET ? ***
```

Fig. 2.8 Boule de neige system at a European table

```
?20


            *** MESSIEURS, FAITES VOS JEUX ...***

?CHANCES SIMPLES
A CHEVAL ? (ENTER YES OR NO) ?NO
ENTER 'ROUGE', OR 'NOIR', OR 'PAIR', OR 'IMPAIR',
OR 'MANQUE', OR 'PASSE' ?MANQUE

        *** LES JEUX SONT FAITS, RIEN NE VA PLUS...***

            *** 7     ,ROUGE ,IMPAIR ET MANQUE ***


        *** CONGRATULATIONS ! YOU JUST WON  20   FRANCS !! ***

YOU NOW HAVE  120  FRANCS. DO YOU WANT TO TRY AGAIN ?YES

                *** HOW MUCH DO YOU WANT TO BET ? ***
?40


            *** MESSIEURS, FAITES VOS JEUX ...***

?CHANCES SIMPLES
A CHEVAL ? (ENTER YES OR NO) ?NO
ENTER 'ROUGE', OR 'NOIR', OR 'PAIR', OR 'IMPAIR',
OR 'MANQUE', OR 'PASSE' ?MANQUE

        *** LES JEUX SONT FAITS, RIEN NE VA PLUS...***

            *** 15    ,NOIR  ,IMPAIR ET MANQUE ***


        *** CONGRATULATIONS ! YOU JUST WON   40   FRANCS !! ***

YOU NOW HAVE  160  FRANCS. DO YOU WANT TO TRY AGAIN ?YES

                *** HOW MUCH DO YOU WANT TO BET ? ***
?80


            *** MESSIEURS, FAITES VOS JEUX ...***

?CHANCES SIMPLES
```

Fig. 2.8 Boule de neige system at a European table (cont'd)

```
A CHEVAL ? (ENTER YES OR NO) ?NO
ENTER 'ROUGE', OR 'NOIR', OR 'PAIR', OR 'IMPAIR',
OR 'MANQUE', OR 'PASSE' ?MANQUE


          *** LES JEUX SONT FAITS, RIEN NE VA PLUS...***


            ***  33    ,NOIR  ,IMPAIR ET PASSE  ***



YOU JUST LOST  80    FRANCS !

YOU NOW HAVE  80    FRANCS. DO YOU WANT TO TRY AGAIN ?YES


            *** HOW MUCH DO YOU WANT TO BET ? ***
?80


            *** MESSIEURS, FAITES VOS JEUX ...***


?CHANCES SIMPLES
A CHEVAL ? (ENTER YES OR NO) ?NO
ENTER 'ROUGE', OR 'NOIR', OR 'PAIR', OR 'IMPAIR',
OR 'MANQUE', OR 'PASSE' ?MANQUE


          *** LES JEUX SONT FAITS, RIEN NE VA PLUS...***


            ***  5     ,ROUGE ,IMPAIR ET MANQUE ***



      *** CONGRATULATIONS ! YOU JUST WON  80    FRANCS !! ***

YOU NOW HAVE  160  FRANCS. DO YOU WANT TO TRY AGAIN ?YES


            *** HOW MUCH DO YOU WANT TO BET ? ***
?160


            *** MESSIEURS, FAITES VOS JEUX ...***


?CHANCES SIMPLES
A CHEVAL ? (ENTER YES OR NO) ?NO
ENTER 'ROUGE', OR 'NOIR', OR 'PAIR', OR 'IMPAIR',
OR 'MANQUE', OR 'PASSE' ?MANQUE


          *** LES JEUX SONT FAITS, RIEN NE VA PLUS...***
```

Fig. 2.8 Boule de neige system at a European table (cont'd)

```
       ***  12   ,ROUGE ,PAIR   ET MANQUE ***


   *** CONGRATULATIONS ! YOU JUST WON  160  FRANCS !! ***

YOU NOW HAVE  320  FRANCS. DO YOU WANT TO TRY AGAIN ?NO

GETTING COLD FEET ? AFTER ALL, IT'S ONLY MONEY !

DONE
```

Fig. 2.8 Boule de neige system at a European table (cont'd)

Ain't we lucky? (Note that we by-passed the introductory explanations by adding the following two statements,

$$181 \quad T=1$$
$$182 \quad GOTO \ 330$$

to the program that is displayed in Fig. 2.14.)

### The Paroli

The *paroli*, another "foolproof" system for getting rich by betting on a *chance simple*, is mildly profitable if and when "runs of four" become common. You wait until a particular *chance simple*, say *noir*, comes up four times in a row. You restrain yourself just a little longer until *rouge* comes up, and then, hoping for a run of three more successive *rouge*, you plunge into the *paroli*. First, you place your basic bet of B francs (20, 50, or whatever) on *rouge*. If you lose, you start all over again. If you win, you bet 3B francs on the next *coup*. If you win, you double your bet to 6B francs on the third *coup*. Remember that only 2B of this sum are actually your own! If you win again, you withdraw the 12B francs for a net gain of 10B francs. If you lose any one coup, you withdraw. The most you can ever lose is 2B francs. You can thus play the *paroli* at least five times and lose at most 10B francs—the amount you can win in just one progression—provided the gods are with you.

We have compiled all the possibilities in Table 2.3.

Let us move on to a discussion of some other "infallible" systems. Xan Fielding, we might add, comments very appropriately that "all are infallible in the regularity with which they have proved failures" (Ref. [3], p. 169).

**Table 2.3   Example of paroli system**

| Coup No. | Bet | Win (W) Lose (L) | Net gain (loss) |
|----------|-----|------------------|-----------------|
| 1 | B  | W | B    |
| 2 | 3B | W | 4B   |
| 3 | 6B | W | 10B  |
| 1 | B  | W | B    |
| 2 | 3B | W | 4B   |
| 3 | 6B | L | (2B) |
| 1 | B  | W | B    |
| 2 | 3B | L | (2B) |
| 1 | B  | L | (B)  |

## Montant d'Alembert (d'Alembert's Progression)

D'Alembert[1] suggested increasing the stake on a *chance simple* after each loss by an amount equal to the first (basic) bet B and to decrease it by that amount after each win. (As basic bet B we may take the house minimum.)

Suppose that we back *pair* and our basic bet is 10 francs. Table 2.4 illustrates what might happen. Because of the rule to reduce the stake after each win by the amount of the basic bet, the bet at *coup* No. 10 would be 0 francs, and the progression comes to a halt. We wind up with a net gain of 50 francs, and we note that we were never down by more than 60 francs.

**Table 2.4   Example of d'Alembert's progression**

| Coup No. | Bet | Result | Win (W) Lose (L) | Net gain (loss) |
|----------|-----|--------|------------------|-----------------|
| 1  | 10 | 13 | L | (10) |
| 2  | 20 | 27 | L | (30) |
| 3  | 30 | 1  | L | (60) |
| 4  | 40 | 2  | W | (20) |
| 5  | 30 | 16 | W | 10  |
| 6  | 20 | 8  | W | 30  |
| 7  | 10 | 0  | L | 20  |
| 8  | 20 | 8  | W | 40  |
| 9  | 10 | 22 | W | 50  |
| 10 | 0  |    |   |     |

Such a d'Alembert progression always comes to a halt after an odd number of turns (1, 3, 5, etc.), and you wind up with a net gain

[1] Jean Le Rond d'Alembert (1717–1783), French mathematician and physicist.

of B francs if it comes to a halt after the first turn, 2B francs if it comes to a halt after the third turn, 3B francs if it comes to a halt after the fifth turn, and so forth.

In general, if your basic bet is B and the progression terminates with coup number $(2n + 1)$, where $n = 0, 1, 2, 3, \ldots$, then

(1)
Net gain $= (n + 1)B$
Largest single stake ever required $= (n + 1)B$
Largest amount ever in the hole $= (B/2)n(n + 1)$

In our example in Table 2.4, the progression terminated with the ninth coup. So, since $2n + 1 = 9$, $n = 4$. If the basic bet $B = 10$, we obtain

Net gain $= 50$
Largest single stake ever required $= 50$
Largest amount ever in the hole $= 5 \times 4 \times 5 = 100$

We were lucky inasmuch as the largest stake we ever had to put up was 40 francs (and not 50 francs) and the most we were ever in the hole was 60 francs (and not 100 francs). Our formulas in (1) pertain to the worst progression in which all n losses occur in the beginning and are followed by $(n + 1)$ consecutive wins.

If you should wish to get out *before* the progression comes to a natural halt—say, after m turns—here is what you should know: The most you could possibly have lost by then is

$$(B/2)m(m + 1)$$

and the biggest possible stake you would have been required to put up was Bm. For example, if you want to quit after five turns, you win either B or 3B francs, or else, you are at most 15B francs in the hole and at no time did you have to put up a stake in excess of 5B francs. Compare this with a martingale progression of five turns. You either win B francs or you are 31B francs in the hole, with 16B francs being the largest stake required. To get out of the hole, if possible, would require a stake of 31B. To drive home the point, let us look at a sequence of ten consecutive losses and calculate at every step the relevant data for a martingale progression and a d'Alembert progression. In Table 2.5, we place these data in juxtaposition.

Assuming that the house maximum is 1,000 times the house minimum (smaller casinos may not offer such a wide spread), you may play a progression of at least 1,000 *coups* with the d'Alembert but only of 10 *coups* with the martingale. Let it be noted in this context that 28 consecutive *noir* were once recorded (Ref. [5], p. 29). Shorter runs of a single color are not at all uncommon. This author has

Table 2.5 Comparison of martingale and d'Alembert progressions

| Coup No. | Win (W) Lose (L) | Bet | | Net loss | |
|---|---|---|---|---|---|
| | | Martingale | d'Alembert | Martingale | d'Alembert |
| 1 | L | 10 | 10 | 10 | 10 |
| 2 | L | 20 | 20 | 30 | 30 |
| 3 | L | 40 | 30 | 70 | 60 |
| 4 | L | 80 | 40 | 150 | 100 |
| 5 | L | 160 | 50 | 310 | 150 |
| 6 | L | 320 | 60 | 630 | 210 |
| 7 | L | 640 | 70 | 1270 | 280 |
| 8 | L | 1280 | 80 | 2550 | 360 |
| 9 | L | 2560 | 90 | 5110 | 450 |
| 10 | L | 5120 | 100 | 10230 | 550 |

witnessed, with considerable discomfort, eight *rouge* in succession. Needless to say, he was backing *noir*.

Let's try the d'Alembert progression on the computer. First, we run a bunch of random numbers (spin the wheel a number of times; we programmed the computer to do this quickly and efficiently by placing bets in the amount 0) until we find that one of the *chances simples* has appeared with greater frequency than the others. As soon as this happens, we embark on the d'Alembert progression, backing the opposite *chance simple* (see Fig. 2.9). (Note that we could even have our program record the relative frequencies of the *chance simples* after each *coup*, or, better still, their deviations from the norm; see also p. 74.)

```
DO YOU WANT TO PLAY AT THE EUROPEAN TABLE OR THE
AMERICAN TABLE ?EUROPEAN


HOW MUCH MONEY DO YOU HAVE ?1000
IN WHAT CURRENCY ?FRANCS



              *** BONNE CHANCE ***



         *** HOW MUCH DO YOU WANT TO BET ? ***
?0
         ***  5     ,ROUGE ,IMPAIR ET MANQUE ***
```

Fig. 2.9 D'Alembert progression at a European table

```
          *** HOW MUCH DO YOU WANT TO BET ? ***
?0
             ***  20    ,NOIR  ,PAIR   ET PASSE  ***


          *** HOW MUCH DO YOU WANT TO BET ? ***
?0
             ***  8     ,NOIR  ,PAIR   ET MANQUE ***


          *** HOW MUCH DO YOU WANT TO BET ? ***
?0
             ***  15   ,NOIR  ,IMPAIR ET MANQUE ***


          *** HOW MUCH DO YOU WANT TO BET ? ***
?10


          *** MESSIEURS, FAITES VOS JEUX ...***


?CHANCES SIMPLES
A CHEVAL ? (ENTER YES OR NO) ?NO
ENTER 'ROUGE', OR 'NOIR', OR 'PAIR', OR 'IMPAIR',
OR 'MANQUE', OR 'PASSE' ?ROUGE


       *** LES JEUX SONT FAITS, RIEN NE VA PLUS...***


          ***  26   ,NOIR  ,PAIR   ET PASSE  ***



YOU JUST LOST  10    FRANCS !

YOU NOW HAVE  990   FRANCS. DO YOU WANT TO TRY AGAIN ?YES


          *** HOW MUCH DO YOU WANT TO BET ? ***
?20


          *** MESSIEURS, FAITES VOS JEUX ...***


?CHANCES SIMPLES
A CHEVAL ? (ENTER YES OR NO) ?NO
ENTER 'ROUGE', OR 'NOIR', OR 'PAIR', OR 'IMPAIR',
OR 'MANQUE', OR 'PASSE' ?ROUGE



       *** LES JEUX SONT FAITS, RIEN NE VA PLUS...***
```

*Fig. 2.9 D'Alembert progression at a European table (cont'd)*

```
                    *** ZERO ***

          *** DO YOU WANT HALF YOUR STAKE BACK ? ***

?NO
                  *** EN PRISON...HERE WE GO AGAIN ***


            ***  19    ,ROUGE ,IMPAIR ET PASSE   ***


YOU BROKE EVEN !
YOU NOW HAVE  990  FRANCS. DO YOU WANT TO TRY AGAIN ?YES


              *** HOW MUCH DO YOU WANT TO BET ? ***
?20


              *** MESSIEURS, FAITES VOS JEUX ...***


?CHANCES SIMPLES
A CHEVAL ? (ENTER YES OR NO) ?NO
ENTER 'ROUGE', OR 'NOIR', OR 'PAIR', OR 'IMPAIR',
OR 'MANQUE', OR 'PASSE' ?ROUGE


        *** LES JEUX SONT FAITS, RIEN NE VA PLUS...***


            ***  11    ,NOIR  ,IMPAIR ET MANQUE ***



YOU JUST LOST  20   FRANCS


YOU NOW HAVE  970  FRANCS. DO YOU WANT TO TRY AGAIN ?YES


              *** HOW MUCH DO YOU WANT TO BET ? ***
?30


              *** MESSIEURS, FAITES VOS JEUX ...***


?CHANCES SIMPLES
A CHEVAL ? (ENTER YES OR NO) ?NO
ENTER 'ROUGE', OR 'NOIR', OR 'PAIR', OR 'IMPAIR',
OR 'MANQUE', OR 'PASSE' ?ROUGE


        *** LES JEUX SONT FAITS, RIEN NE VA PLUS...***


            ***  30    ,ROUGE ,PAIR   ET PASSE  ***
```
Fig. 2.9 *D'Alembert progression at a European table (cont'd)*

```
        *** CONGRATULATIONS ! YOU JUST WON  30    FRANCS !! ***

YOU NOW HAVE  1000     FRANCS. DO YOU WANT TO TRY AGAIN ?YES


                *** HOW MUCH DO YOU WANT TO BET ? ***
?20



                *** MESSIEURS, FAITES VOS JEUX ...***

?CHANCES SIMPLES
A CHEVAL ? (ENTER YES OR NO) ?NO
ENTER 'ROUGE', OR 'NOIR', OR 'PAIR', OR 'IMPAIR',
OR 'MANQUE', OR 'PASSE' ?ROUGE


           *** LES JEUX SONT FAITS, RIEN NE VA PLUS...***


           ***  24     ,NOIR  ,PAIR    ET PASSE  ***



YOU JUST LOST  20    FRANCS !

YOU NOW HAVE  980  FRANCS. DO YOU WANT TO TRY AGAIN ?YES


                *** HOW MUCH DO YOU WANT TO BET ? ***
?30


                *** MESSIEURS, FAITES VOS JEUX ...***


?CHANCES SIMPLES
A CHEVAL ? (ENTER YES OR NO) ?NO
ENTER 'ROUGE', OR 'NOIR', OR 'PAIR', OR 'IMPAIR',
OR 'MANQUE', OR 'PASSE' ?ROUGE


           *** LES JEUX SONT FAITS, RIEN NE VA PLUS...***


           ***  3      ,ROUGE ,IMPAIR ET MANQUE ***



        *** CONGRATULATIONS ! YOU JUST WON  30    FRANCS !! ***

YOU NOW HAVE  1010     FRANCS. DO YOU WANT TO TRY AGAIN ?YES
```

*Fig. 2.9 D'Alembert progression at a European table (cont'd)*

```
          *** HOW MUCH DO YOU WANT TO BET ? ***
?20


          *** MESSIEURS, FAITES VOS JEUX ...***

?CHANCES SIMPLES
A CHEVAL ? (ENTER YES OR NO) ?NO
ENTER 'ROUGE', OR 'NOIR', OR 'PAIR', OR 'IMPAIR',
OR 'MANQUE', OR 'PASSE' ?ROUGE

       *** LES JEUX SONT FAITS, RIEN NE VA PLUS...***

        ***   18    ,ROUGE ,PAIR   ET MANQUE ***


    *** CONGRATULATIONS ! YOU JUST WON  20   FRANCS !! ***

YOU NOW HAVE  1030    FRANCS. DO YOU WANT TO TRY AGAIN ?YES



          *** HOW MUCH DO YOU WANT TO BET ? ***
?10


          *** MESSIEURS, FAITES VOS JEUX ...***

?CHANCES SIMPLES
A CHEVAL ? (ENTER YES OR NO) ?NO
ENTER 'ROUGE', OR 'NOIR', OR 'PAIR', OR 'IMPAIR',
OR 'MANQUE', OR 'PASSE' ?ROUGE

       *** LES JEUX SONT FAITS, RIEN NE VA PLUS...***

        ***   34    ,ROUGE ,PAIR   ET PASSE  ***


    *** CONGRATULATIONS ! YOU JUST WON  10   FRANCS !! ***

YOU NOW HAVE  1040    FRANCS. DO YOU WANT TO TRY AGAIN ?NO

GETTING COLD FEET ? AFTER ALL, IT'S ONLY MONEY !
DONE
```

Fig. 2.9 D'Alembert progression at a European table (cont'd)

Figure 2.9 reveals that we managed to win a grand total of 40 francs with 13 spins of the wheel, which should take about 26 minutes in Monte Carlo. This works out to approximately $20 per hour, or nothing to get excited about.

Note (in Fig. 2.9) that, when the 0 came up, we decided to let our stake ride for the next *coup*. Since we won the next *coup*, we broke even and hence could ignore both *coups* because nothing had really happened. We picked up the d'Alembert progression where we left it *before* the zero came up. (Had we lost the *coup* after the zero, it would have counted as an ordinary loss, and we would have augmented our stake by 10 francs for the next *coup*.)

## Progression on the Double Dozen

In this system, you wait until the first or the last dozen comes up three or four times in a row, and then you place a split bet on the other two. If you lose, you triple your bet; if you win, you get out.

If you place such a bet at the European table, we recommend that you put either one-quarter of your stake on the *sixain* 19–20–21–22–23–24 and the remaining three-quarters on *manque* (low) for a bet on the first and second dozen, or on the *sixain* 13–14–15–16–17–18 and on *passe* (high) for a bet on the second and the third dozen. If a zero comes up, you have a 50 percent chance of retaining three-fourths of your stake rather than losing it all. The same method of staking is recomn ided when you play at an American table when half your stake is returned on all 1–1 "outside" bets on a 0 and a 00.

Suppose that we start with 40 francs. Table 2.6 illustrates what might happen when you back the second and third dozen.

Table 2.6    Progression on the double dozen

| Coup No. | Bet | Result | Win (W) Lose (L) | Net gain (loss) |
|---|---|---|---|---|
| 1 | 40 | 8 | L | (40) |
| 2 | 120 | 3 | L | (160) |
| 3 | 360 | 0 | L | (520) |
| 4 | 1080 | 5 | L | (1600) |
| 5 | 3240 | 27 | W | 20 |

Since the losses suffered during the first n turns, assuming a basic bet B, amount to

$$B + 3B + 9B + \ldots + 3^{n-1}B = B(1 + 3 + \ldots + 3^{n-1}) = B\frac{1 - 3^n}{1 - 3}$$

and since you win one-half of your stake $3^n$ B, which you have to bet at the $(n + 1)$th turn—if you win at that turn—your net gain, as soon as you win, is always

$$\frac{3^n B}{2} - B \frac{1 - 3^n}{1 - 3} = \frac{B}{2}$$

(20 francs in our case). *However*, this result is valid only as long as you can triple your bet. If the house maximum is 1000 times the minimum and you start out with the minimum bet, then the system collapses as soon as $3^n$ exceeds 1000. Since $3^6 = 729$ and $3^7 = 2187$, the system collapses after seven consecutive losses, and you are 1093B francs in the hole. Even if you win the next *coup*, the most you can recoup is 500B francs!

The same system applies to columns as well as dozens.

Let's see how a progression on the double dozen goes on the computer. Again, we run our program without betting until the first or the third dozen has come up three times in a row. Then we plunge ahead and back the other two dozen. The run is shown in Fig. 2.10.

```
DO YOU WANT TO PLAY AT THE EUROPEAN TABLE OR THE
AMERICAN TABLE ?AMERICAN

HOW MUCH MONEY DO YOU HAVE ?1000

                    *** GOOD LUCK ***

          *** HOW MUCH DO YOU WANT TO BET ? ***
?O
          ***   29   ,BLACK ,ODD    AND HIGH    ***

          *** HOW MUCH DO YOU WANT TO BET ? ***
?O
          ***   13   ,BLACK ,ODD    AND LOW     ***

          *** HOW MUCH DO YOU WANT TO BET ? ***
?O
          ***   2    ,BLACK ,EVEN   AND LOW     ***

          *** HOW MUCH DO YOU WANT TO BET ? ***
?O
```

Fig. 2.10 Progression on the double dozen at an American table

```
        ***  10    ,BLACK ,EVEN   AND LOW    ***


          *** HOW MUCH DO YOU WANT TO BET ? ***
?0
          ***  7     ,RED   ,ODD    AND LOW    ***


          *** HOW MUCH DO YOU WANT TO BET ? ***
?20


          *** MESSIEURS, FAITES VOS JEUX ...***

?DOZEN
SPLIT ? (ENTER YES OR NO) ?YES
WHICH DOZEN ?
ENTER 'FIRST', 'SECOND', OR 'THIRD' ?SECOND
AND WHICH ADJACENT DOZEN ?THIRD


        *** LES JEUX SONT FAITS, RIEN NE VA PLUS...***


                  *** DOUBLE ZERO ***


YOU JUST LOST  20    DOLLARS !

YOU NOW HAVE  980   DOLLARS. DO YOU WANT TO TRY AGAIN ?YES


          *** HOW MUCH DO YOU WANT TO BET ? ***
?60


          *** MESSIEURS, FAITES VOS JEUX ...***

?DOZEN
SPLIT ? (ENTER YES OR NO) ?YES
WHICH DOZEN ?
ENTER 'FIRST', 'SECOND', OR 'THIRD' ?SECOND
AND WHICH ADJACENT DOZEN ?THIRD


        *** LES JEUX SONT FAITS, RIEN NE VA PLUS...***


        ***  12    ,RED   ,EVEN   AND LOW    ***


YOU JUST LOST  60    DOLLARS !
```

Fig. 2.10 Progression on the double dozen at an American table (cont'd)

```
YOU NOW HAVE  920   DOLLARS. DO YOU WANT TO TRY AGAIN ?YES


            *** HOW MUCH DO YOU WANT TO BET ? ***

?180


            *** MESSIEURS, FAITES VOS JEUX ...***


?DOZEN
SPLIT ? (ENTER YES OR NO) ?YES
WHICH DOZEN ?
ENTER 'FIRST', 'SECOND', OR 'THIRD' ?SECOND
AND WHICH ADJACENT DOZEN ?THIRD


        *** LES JEUX SONT FAITS, RIEN NE VA PLUS...***


            *** 9     ,RED  ,ODD    AND LOW    ***


YOU JUST LOST  180   DOLLARS !


YOU NOW HAVE  740   DOLLARS. DO YOU WANT TO TRY AGAIN ?YES


            *** HOW MUCH DO YOU WANT TO BET ? ***
?540


            *** MESSIEURS, FAITES VOS JEUX ...***


?DOZEN
SPLIT ? (ENTER YES OR NO) ?YES
WHICH DOZEN ?
ENTER 'FIRST', 'SECOND', OR 'THIRD' ?SECOND
AND WHICH ADJACENT DOZEN ?THIRD


        *** LES JEUX SONT FAITS, RIEN NE VA PLUS...***


            *** 23    ,RED  ,ODD    AND HIGH   ***


     *** CONGRATULATIONS ! YOU JUST WON  270   DOLLARS !! ***


YOU NOW HAVE  1010    DOLLARS. DO YOU WANT TO TRY AGAIN ?NO


GETTING COLD FEET ? AFTER ALL, IT'S ONLY MONEY !

DONE
```

Fig. 2.10 Progression on the double dozen at an American table (cont'd)

Figure 2.10 reveals that we squeezed out $10 after nine spins of the wheel. In Las Vegas, this would take about five minutes, and so we would appear to have made money at the rate of $120 per hour! Don't kid yourself, however; it usually does not work out this way. (Had we played with a basic stake of only 10 francs at Monte Carlo, our earnings would have been less than $10 per hour.)

The reader may let his imagination be his guide and invent his own system. What do you have to lose as long as you play on the computer? Besides, if you can't make money playing your system, you can always try to sell it! Bob Martin has said, "The only sure way to make money on any system . . . is to sell it to other gamblers" (Ref. [7], p. 91).

## THE COMPUTER PROGRAM

The game of roulette may be simulated on a computer at many different levels of sophistication. Basically, all we really need the computer for is to feed us numbers 0, 1, 2, 3, . . . , 36 (00, 0, 1, 2, 3, . . . 36 at the American table), at random. By doing so, the computer assumes the function of the wheel.

Figure 2.11 provides a very rudimentary simulation. Note how line 60 produces the numbers 0, 1, 2, 3, . . . , 36, each with probability 1/37, and how line 80 produces the numbers −1, 0, 1, 2, 3, . . . , 36, each with probability 1/38. Since the computer does not recognize 00 as a number, we let −1 play the role of 00 until it comes to the print-out. At that point, we represent it by the character string "00".

```
10    DIM T$[8]
20    PRINT "DO YOU WANT TO PLAY AT THE EUROPEAN TABLE OR"
30    PRINT "AT THE AMERICAN TABLE ";
40    INPUT T$
50    IF T$[1,1]="A" THEN 80
60    U=INT(RND(1)*37)
70    GOTO 100
80    U=INT(RND(1)*38)-1
90    IF U=-1 THEN 120
100   PRINT U
110   GOTO 130
120   PRINT "00"
130   PRINT "DO YOU WANT TO PLAY AGAIN ";
140   INPUT T$
150   IF T$[1,1]="Y" THEN 20
160   END
```

Fig. 2.11 Rudimentary roulette program

The program in Fig. 2.11 provides the bare essentials only. You will have to do your own conversions of the results into *noir* or *rouge*, *colonne* or *douzaine*, etc., and you will have to do your own bookkeeping. Also note that this program does not provide for the concomitant ballyhoo that would simulate the proper gambling atmosphere.

As a next step, we will (1) classify the result by color, even or odd, high or low, (2) have the computer accept bets on *chances simples,* (3) have the computer decide whether a bet won or lost, and (4) throw in some atmosphere by programming the computer to simulate the *croupier's* incantations. This improved program is shown in Fig. 2.12.

```
10    DIM T$[8],A$[6],B$[6],C$[6],F$[8],G$[12],K$[36]
20    PRINT "DO YOU WANT TO PLAY AT THE EUROPEAN TABLE OR AT"
30    PRINT "THE AMERICAN TABLE ";
40    INPUT T$
50    IF T$[1,1]="A" THEN 90
60    T=1
70    U=INT(RND(1)*37)
80    GOTO 110
90    U=INT(RND(1)*38)-1
100    T=2
110    C2=SGN(U/2-INT(U/2))+3
120    C3=6-INT((U-1)/18)
130    IF (U/2)=INT(U/2) AND (U<11 OR (U>19 AND U<29)) THEN 170
140    IF (U/2 <> INT(U/2)) AND ((U>10 AND U<18) OR U>28) THEN 170
150    C1=2
160    GOTO 180
170    C1=1
180    K$="BLACK RED    EVEN   ODD    HIGH   LOW
190    A$=K$[(C1-1)*6+1,(C1-1)*6+6]
200    B$=K$[(C2-3)*6+13,(C2-3)*6+18]
210    C$=K$[(C3-5)*6+25,(C3-5)*6+30]
220    IF T=3 THEN 330
230    PRINT TAB(13)"*** MESSIEURS, FAITES VOS JEUX ***"
240    PRINT "ENTER 'BLACK', OR 'RED', OR 'EVEN', OR 'ODD',"
250    PRINT "OR 'HIGH', OR 'LOW' ";
260    INPUT F$
270    G$="LAEDVEDDIGOW"
280    FOR J=0 TO 5
290    IF F$[2,3]=G$[1+2*J,2+2*J] THEN 310
300    NEXT J
310    PRINT TAB(8)"*** LES JEUX SONT FAITS, RIEN NE VA ";
320    PRINT "PLUS ...***"
330    IF U=0 THEN 470
340    IF U=-1 THEN 520
350    PRINT TAB(12)"*** "U","A$","B$" AND "C$" ***"
360    IF J+1=C1 OR J+1=C2 OR J+1=C3 THEN 390
370    PRINT "YOU LOST !"
380    GOTO 430
390    IF T=3 THEN 420
400    PRINT "YOU WON !"
410    GOTO 430
420    PRINT "YOU BROKE EVEN !"
430    PRINT "DO YOU WANT TO PLAY AGAIN ";
440    INPUT F$
450    IF F$[1,1]="N" THEN 540
460    GOTO 20
470    PRINT TAB(25)"*** ZERO ***"
```

Fig. 2.12 Improved roulette program

```
480   IF T >= 2 THEN 370
490   T=3
500   PRINT TAB(13)"*** EN PRISON...HERE WE GO AGAIN ***"
510   GOTO 70
520   PRINT TAB(21)"*** DOUBLE ZERO ***"
530   GOTO 370
540   END
```

*Fig. 2.12 Improved roulette program (cont'd)*

Note how we introduced a numerical code for *noir, rouge, passe,* etc., in lines 110, 120, 150, and 170, and how we translated the code into words by using K$ in lines 190 to 210.

Note also how in 480 we sent the player at the American table to the "YOU LOST!" message after a zero. To take care of the player at the European table, which allows prison of one degree only and does not give the player the option of getting half his stake back, we use the "table variable T" and set it to 3 in line 490. (Remember that the "T = 2" folks are already out of the way.) We pass control to line 70 but, to circumvent the betting, the *croupier's* chant, and the rest of it, we go directly from line 210 to the print-out of the result of the next *coup.* If this should be another zero, then the statement in line 480 will take care of passing control to the "YOU LOST!" message. The statement in line 390 takes care of converting a "win" to a "break-even" after winning a *coup* when *en prison.*

We display a sample run of this program in Fig. 2.13.

```
DO YOU WANT TO PLAY AT THE EUROPEAN TABLE OR AT
THE AMERICAN TABLE ?AMERICAN
                *** MESSIEURS, FAITES VOS JEUX ***
ENTER 'BLACK', OR 'RED', OR 'EVEN', OR 'ODD',
OR 'HIGH', OR 'LOW' ?BLACK
            *** LES JEUX SONT FAITS, RIEN NE VA PLUS ...***
            ***  15   ,BLACK ,ODD    AND LOW    ***
YOU WON !
DO YOU WANT TO PLAY AGAIN ?YES
DO YOU WANT TO PLAY AT THE EUROPEAN TABLE OR AT
THE AMERICAN TABLE ?AMERICAN
                *** MESSIEURS, FAITES VOS JEUX ***
ENTER 'BLACK', OR 'RED', OR 'EVEN', OR 'ODD',
OR 'HIGH', OR 'LOW' ?RED
            *** LES JEUX SONT FAITS, RIEN NE VA PLUS ...***
            ***  35   ,BLACK ,ODD    AND HIGH   ***
YOU LOST !
DO YOU WANT TO PLAY AGAIN ?YES
DO YOU WANT TO PLAY AT THE EUROPEAN TABLE OR AT
THE AMERICAN TABLE ?AMERICAN
                *** MESSIEURS, FAITES VOS JEUX ***
ENTER 'BLACK', OR 'RED', OR 'EVEN', OR 'ODD',
OR 'HIGH', OR 'LOW' ?EVEN
            *** LES JEUX SONT FAITS, RIEN NE VA PLUS ...***
            ***  30   ,RED   ,EVEN   AND HIGH   ***
YOU WON !
```

*Fig. 2.13 Sample run of improved roulette program*

```
DO YOU WANT TO PLAY AGAIN ?YES
DO YOU WANT TO PLAY AT THE EUROPEAN TABLE OR AT
THE AMERICAN TABLE ?EUROPEAN
                *** MESSIEURS, FAITES VOS JEUX ***
ENTER 'BLACK', OR 'RED', OR 'EVEN', OR 'ODD',
OR 'HIGH', OR 'LOW' ?ODD
          *** LES JEUX SONT FAITS, RIEN NE VA PLUS ...***
            ***  13  ,BLACK ,ODD    AND LOW   ***
YOU WON !
DO YOU WANT TO PLAY AGAIN ?YES
DO YOU WANT TO PLAY AT THE EUROPEAN TABLE OR AT
THE AMERICAN TABLE ?EUROPEAN
                *** MESSIEURS, FAITES VOS JEUX ***
ENTER 'BLACK', OR 'RED', OR 'EVEN', OR 'ODD',
OR 'HIGH', OR 'LOW' ?HIGH
          *** LES JEUX SONT FAITS, RIEN NE VA PLUS ...***
            ***  24  ,BLACK ,EVEN   AND HIGH   ***
YOU WON !
DO YOU WANT TO PLAY AGAIN ?YES
DO YOU WANT TO PLAY AT THE EUROPEAN TABLE OR AT
THE AMERICAN TABLE ?EUROPEAN
                *** MESSIEURS, FAITES VOS JEUX ***
ENTER 'BLACK', OR 'RED', OR 'EVEN', OR 'ODD',
OR 'HIGH', OR 'LOW' ?LOW
          *** LES JEUX SONT FAITS, RIEN NE VA PLUS ...***
                    *** ZERO ***
            *** EN PRISON...HERE WE GO AGAIN ***
                    *** ZERO ***
YOU LOST !
DO YOU WANT TO PLAY AGAIN ?NO

DONE
```

Fig. 2.13 Sample run of improved roulette program (cont'd)

Since our computer does not allow the use of the logical operators AND and OR in conjunction with string statements, we had to convert input *NOIR, ROUGE, PAIR, IMPAIR, PASSE,* and *MANQUE* into numbers so that the computer could decide in line 360 whether the player lost or won. The conversion was effected by means of loop 280–300, using the second and third letters of the input for identification. (It would have been simpler to use the first letters only, but there is no way of keeping the six possible inputs apart in the French version by using only one letter in a fixed position—NRP̲IP̲M, OO̲A-MA̲A̲, I̲U̲IPSN, R̲G̲R̲ASQ—nor can it be done by using just the first and second letters—NOR̲O̲P̲A̲IMPA̲MA—and we want to treat both versions simultaneously in our ultimate program.) With this conversion out of the way, it is now an easy matter to compare, in line 360, the input with the result of the *coup.*

Even with this more sophisticated program, you still have to do your own bookkeeping, and if you want to bet on anything other than an even chance, you are strictly on your own.

The program that produced the print-out in Fig. 2.1 (and the ones in Figs. 2.7 to 2.10) not only plays the role of the wheel, but it also assumes the function of the table, the *croupier,* and the *chef de partie* (the headman in charge of the table). It is the player's bookkeeper and his conscience. It keeps records of all bets, rejects inadmissible bets—such as an *à cheval* bet on NOIR-MANQUE or a bet on the (nonexistent) fifth column—keeps track of the player's holdings, rejects bets the player cannot cover, and makes it impossible for the player to go back on his word once he is committed. It also rakes in the losses relentlessly. This master program is shown in Fig. 2.14.

```
10     PRINT TAB(26)"**********"
20     PRINT TAB(26)"*ROULETTE*"
30     PRINT TAB(26)"**********"
40     PRINT LIN(1)
50     DIM A$[6],B$[6],C$[6],D$[16],E$[16],F$[8],G$[12],H$[6]
60     DIM I$[8],K$[41],M$[10]
70     PRINT "DO YOU WANT TO PLAY AT THE EUROPEAN TABLE OR THE "
80     PRINT "AMERICAN TABLE ";
90     INPUT G$
100    PRINT LIN(1)
110    PRINT "HOW MUCH MONEY DO YOU HAVE ";
120    INPUT M
130    IF G$[1,1]="E" THEN 160
140    M$="DOLLARS"
150    GOTO 180
160    PRINT "IN WHAT CURRENCY ";
170    INPUT M$
180    PRINT LIN(1)
190    PRINT "WHEN ASKED TO PLACE YOUR BETS ('MESSIEURS, FAITES";
200    PRINT " VOS JEUX'),"
210    PRINT "ENTER ONE OF THE FOLLOWING:"
220    PRINT LIN(1)
230    IF G$[1,1]="A" THEN 390
235    REM T IS THE 'TABLE VARIABLE'. IT IS 1 FOR THE EUROPEAN
236    REM TABLE AND 2 FOR THE AMERICAN TABLE.
240    T=1
250    PRINT TAB(15)"CHANCES SIMPLES (PAY-OFF: 1 TO 1)"
260    PRINT TAB(15)"EN PLEIN        (PAY-OFF:35 TO 1)"
270    PRINT TAB(15)"COLONNE         (PAY-OFF: 2 TO 1)"
280    PRINT TAB(15)"DOUZAINE        (PAY-OFF: 2 TO 1)"
290    PRINT TAB(15)"TRANSVERSALE    (PAY-OFF:11 TO 1)"
300    PRINT TAB(15)"CARRE           (PAY-OFF: 8 TO 1)"
310    PRINT TAB(15)"SIXAIN          (PAY-OFF: 5 TO 1)"
320    PRINT TAB(15)"QUATRE PREMIERS (PAY-OFF: 8 TO 1)"
330    E$="CECDTCSQNPOZNRAT"
340    K$="NOIR   ROUGE PAIR   IMPAIRPASSE MANQUE ET   "
350    PRINT LIN(1)
360    PRINT TAB(21)"*** BONNE CHANCE ***"
370    PRINT LIN(1)
380    GOTO 530
390    T=2
400    PRINT TAB(15)"EVEN CHANCE     (PAY-OFF: 1 TO 1)"
410    PRINT TAB(15)"STRAIGHT UP     (PAY-OFF:35 TO 1)"
420    PRINT TAB(15)"COLUMN          (PAY-OFF: 2 TO 1)"
430    PRINT TAB(15)"DOZEN           (PAY-OFF: 2 TO 1)"
440    PRINT TAB(15)"STREET          (PAY-OFF:11 TO 1)"
```

Fig. 2.14 Master program "Roulette"

```
450   PRINT TAB(15)"CORNER           (PAY-OFF: 8 TO 1)"
460   PRINT TAB(15)"SIXLINE          (PAY-OFF: 5 TO 1)"
470   PRINT TAB(15)"FIRST FIVE       (PAY-OFF: 6 TO 1)"
480   E$="ESCDSCSFNAUEENLS"
490   K$="BLACK RED   EVEN  ODD   HIGH  LOW     AND "
500   PRINT LIN(1)
510   PRINT TAB(23)"*** GOOD LUCK ***"
520   PRINT LIN(1)
530   P=0
535   REM P, NORMALLY 0, IS SET TO 1 AFTER A ZERO OR A COUP
536   REM NEUTRE.
540   GOTO T OF 550,570
545   REM 550 IS THE OUTCOME OF THE SPIN ON THE EUROPEAN WHEEL
546   REM AND 570 IS THE OUTCOME ON THE AMERICAN WHEEL.
550   U=INT(RND(1)*37)
560   GOTO 580
570   U=INT(RND(1)*38)-1
575   REM C1,C2,C3 ARE THE NUMERICAL CODES FOR BLACK, RED,EVEN,
576   REM ODD, HIGH, LOW.
580   C2=SGN(U/2-INT(U/2))+3
590   C3=6-INT((U-1)/18)
600   IF (U/2=INT(U/2)) AND (U<11 OR (U>19 AND U<29)) THEN 640
610   IF (U/2 <> INT(U/2)) AND ((U>10 AND U<18) OR U>28) THEN 640
620   C1=2
630   GOTO 650
640   C1=1
645   REM A$,B$,C$ EXPRESS THE RESULT IN WORDS SUCH AS 'RED,
646   REM EVEN, LOW'.
650   A$=K$[(C1-1)*6+1,(C1-1)*6+6]
660   B$=K$[(C2-3)*6+13,(C2-3)*6+18]
670   C$=K$[(C3-5)*6+25,(C3-5)*6+30]
680   IF P=1 THEN 1180
690   PRINT LIN(1)
700   PRINT TAB(13)"*** HOW MUCH DO YOU WANT TO BET ? ***"
710   PRINT
720   INPUT B
730   GOTO SGN(B)+1 OF 2820,740
740   IF B>M THEN 700
750   PRINT LIN(1)
760   PRINT TAB(13)"*** MESSIEURS, FAITES VOS JEUX ...***"
770   PRINT LIN(1)
780   INPUT D$
785   REM THE LOOP 790 TO 830 ENCODES THE INPUT IN 780 AND
786   REM ASSIGNS THE VALUE G TO IT.
790   FOR J=1 TO 8
800   IF D$[1,1] <> E$[J,J] THEN 830
810   IF D$[4,4] <> E$[J+8,J+8] THEN 830
820   GOTO 850
830   NEXT J
840   IF J>8 THEN 760
850   G=J
860   IF G>4 OR (G=1 AND T=2) THEN 950
870   GOTO T OF 880,900
880   PRINT "A CHEVAL ? (ENTER YES OR NO) ";
890   GOTO 910
900   PRINT "SPLIT ? (ENTER YES OR NO) ";
910   INPUT F$
915   REM H=1 FOR AN ORDINARY BERT, H=2 FOR A SPLIT BET.
920   H=2-ABS(LEN(F$)-3)
930   IF H>3 OR H<1 THEN 870
940   GOTO 960
```

Fig. 2.14 Master program "Roulette" (cont'd)

```
950   H=1
960   GOTO G OF 970,1280,1590,1770,2040,2360,2440,2520
965   REM 970 TO 1270   DEALS WITH BETS ON AN EVEN CHANCE.
970   IF T=2 THEN 1020
980   PRINT "ENTER 'ROUGE', OR 'NOIR', OR 'PAIR', OR 'IMPAIR',"
990   PRINT "OR 'MANQUE', OR 'PASSE' ";
1000  G$="OIOUAIMPASAN"
1010  GOTO 1050
1020  PRINT "ENTER 'RED', OR 'BLACK', OR 'EVEN', OR 'ODD',"
1030  PRINT "OR 'LOW', OR 'HIGH' ";
1040  G$="LAEDVEDDIGOW"
1050  INPUT F$
1055  REM LOOP 1060 TO 1080 ENCODES THE INPUT IN 1050
1060  FOR J=0 TO 5
1070  IF F$[2,3]=G$[1+2*J,2+2*J] THEN 1090
1080  NEXT J
1090  IF J>5 THEN 970
1100  IF H=1 THEN 1180
1110  PRINT "AND ? (ENTER AN ADJACENT 'CHANCE SIMPLE')"
1120  INPUT I$
1130  FOR L=0 TO 5
1140  IF I$[2,3]=G$[1+2*L,2+2*L] THEN 1160
1150  NEXT L
1160  IF L>5 THEN 1110
1165  REM IN 1170, THE COMPUTER CHECKS THE LEGITIMACY
1166  REM OF A SPLIT BET.
1170  IF (J+L)/2 <> INT((J+L)/2) OR ABS(J-L) <> 2 THEN 1110
1180  GOSUB 2770
1190  W=B-P*B+(H-1)*B*P
1195  REM ON K=0 YOU LOSE A SPLIT BET, ON K=1 YOU BREAK
1196  REM EVEN, ON K=2 YOU WIN.
1200  K=0
1210  IF J+1 <> C1 AND J+1 <> C2 AND J+1 <> C3 THEN 1240
1220  K=K+1
1230  GOTO H OF 2680,1250
1240  GOTO H OF 2590,1250
1250  IF L+1 <> C1 AND L+1 <> C2 AND L+1 <> C3 THEN 1270
1260  K=K+1
1270  GOTO K+1 OF 2590,2560,2710
1275  REM 1280 TO 1580 DEALS WITH BETS ON SINGLE NUMBERS.
1280  GOTO T OF 1350,1290
1290  PRINT "DO YOU WANT TO BACK THE 00 ";
1300  INPUT G$
1310  PRINT
1320  IF G$[1,1]="N" THEN 1350
1330  X=-1
1340  GOTO 1380
1350  PRINT "ENTER ONE OF THE NUMBERS 0,1,2,3,...,36 ";
1360  INPUT X
1370  IF X<0 OR X>36 THEN 1350
1380  IF H=1 THEN 1510
1390  PRINT "AND AN ADJACENT NUMBER ";
1400  INPUT Y
1405  REM IN 1410 TO 1500, THE COMPUTER CHECKS THE LEGITI-
1406  REM MACY OF A SPLIT BET.
1410  IF X=-1 AND (Y=2 OR Y=3 OR Y=0) THEN 1510
1420  IF X=0 AND (Y=1 OR Y=2) THEN 1510
1430  IF Y=0 AND (X=1 OR X=2) THEN 1510
1440  IF (X-1)/3=INT((X-1)/3) AND (ABS(X-Y)=3 OR Y=X+1) THEN 1510
1450  IF (X-2)/3 <> INT((X-2)/3) THEN 1480
1460  IF ABS(X-Y) <> 3 AND ABS(X-Y) <> 1 THEN 1480
```

*Fig. 2.14 Master program "Roulette" (cont'd)*

```
1470    GOTO 1510
1480    IF X/3=INT(X/3) AND (ABS(X-Y)=3 OR Y=X-1) THEN 1500
1490    GOTO 1390
1500    IF T=2 AND (X=0 OR Y=0) THEN 1390
1510    GOSUB 2770
1520    IF H=2 THEN 1550
1530    W=35*B
1540    GOTO 1570
1550    W=17*B
1560    IF X=U OR Y=U THEN 2710
1570    IF X=U THEN 2710
1580    GOTO 2590
1585    REM 1590 TO 1760 DEALS WITH BETS AND SPLIT BETS ON
1586    REM COLUMNS.
1590    PRINT "WHICH COLUMN ? (ENTER 1, 2, OR 3) ";
1600    INPUT X
1610    IF X<1 OR X>3 THEN 1590
1620    IF H=1 THEN 1680
1630    PRINT "AND WHICH ADJACENT COLUMN ";
1640    INPUT Y
1645    REM 1650 TO 1670 CHECKS THE LEGITIMACY OF A SPLIT
1646    REM BET ON TWO COLUMNS.
1650    IF (X=1 AND Y=2) OR (X=2 AND (Y=1 OR Y=3)) THEN 1680
1660    IF X=3 AND Y=2 THEN 1680
1670    GOTO 1630
1680    GOSUB 2770
1690    IF H=2 THEN 1720
1700    W=2*B
1710    GOTO 1750
1720    W=B/2
1730    IF X=U-3*INT((U-1)/3) OR Y=U-3*INT((U-1)/3) THEN 2710
1740    GOTO 2590
1750    IF X=U-3*INT((U-1)/3) THEN 2710
1760    GOTO 2590
1765    REM 1770 TO 2030 DEALS WITH BETS AND SPLIT BETS
1766    REM ON DOZENS.
1770    PRINT "WHICH DOZEN ?"
1780    GOTO T OF 1790,1810
1790    PRINT "ENTER 'PREMIERE', MOYENNE', OR 'DERNIERE' ";
1800    GOTO 1820
1810    PRINT "ENTER 'FIRST', 'SECOND', OR 'THIRD' ";
1820    INPUT F$
1830    H$="PMDFST"
1835    REM THE LOOP 1840 TO 1860 ENCODES THE VERBAL INPUT
1836    REM IN LINE 1820
1840    FOR J=1 TO 3
1850    IF F$[1,1]=H$[J+3*(T-1),J+3*(T-1)] THEN 1870
1860    NEXT J
1870    IF J>3 THEN 1770
1880    IF H=1 THEN 2000
1890    PRINT "AND WHICH ADJACENT DOZEN ";
1900    INPUT I$
1910    FOR L=1 TO 3
1920    IF I$[1,1]=H$[L+3*(T-1),L+3*(T-1)] THEN 1940
1930    NEXT L
1940    IF L>3 THEN 1890
1945    REM 1950 CHECKS THE LEGITIMACY OF A SPLIT BET ON
1946    REM TWO ADJACENT DOZENS.
1950    IF ABS(J-L) <> 1 THEN 1890
1960    GOSUB 2770
1970    W=B/2
1980    IF J=INT((U-1)/12)+1 OR L=INT((U-1)/12)+1 THEN 2710
```

Fig. 2.14 Master program "Roulette" (cont'd)

```
1990    GOTO 2590
2000    GOSUB 2770
2010    W=2*B
2020    L=J
2030    GOTO 1980
2035    REM 2040 TO 2350 DEALS WITH 'STREET' BETS.
2040    GOTO T OF 2130,2050
2050    PRINT "DO YOU WANT TO BACK A ROW WITH THE 00 ";
2060    INPUT F$
2070    IF F$[1,1]="N" THEN 2130
2080    PRINT "WITH 0-2 OR WITH 0-3 ? (ENTER 2 OR 3) ";
2090    INPUT Y
2100    IF Y<2 OR Y>3 THEN 2080
2110    Z=-1
2120    GOTO 2280
2130    PRINT "WHICH ROW DO YOU WANT TO BACK ?"
2140    PRINT "ENTER 0, OR 1, OR 2, ..., OR 12 ";
2150    INPUT X
2160    IF X<0 OR X>12 THEN 2130
2170    IF X <> 0 THEN 2320
2180    GOTO T OF 2220,2190
2190    Y=1
2200    Z=2
2210    GOTO 2280
2220    PRINT "WITH 1-2, OR WITH 2-3 ?"
2230    PRINT "ENTER THE TWO NUMBERS, ONE AT A TIME ";
2240    INPUT Y
2250    IF Y<1 OR Y>2 THEN 2220
2260    INPUT Z
2270    IF (Y=1 AND Z <> 2) OR (Y=2 AND Z <> 3) THEN 2220
2280    GOSUB 2770
2290    W=11*B
2300    IF U=0 OR U=Y OR U=Z THEN 2710
2310    GOTO 2590
2320    GOSUB 2770
2330    W=11*B
2340    IF X=INT((U-1)/3)+1 THEN 2710
2350    GOTO 2590
2355    REM 2360 TO 2430 DEALS WITH CORNER BETS.
2360    PRINT "WHICH SQUARE ?"
2370    PRINT "ENTER THE NUMBER FROM THE LEFT UPPER CORNER ";
2380    INPUT X
2390    IF X<1 OR X>32 OR X/3=INT(X/3) THEN 2360
2400    GOSUB 2770
2410    W=8*B
2420    IF U=X OR U=X+1 OR U=X+3 OR U=X+4 THEN 2710
2430    GOTO 2590
2435    REM 2440 TO 2510 DEALS WITH 'SIXLINE' BETS.
2440    PRINT "WHICH IS THE FIRST OF THE TWO ADJACENT ROWS ?"
2450    PRINT "ENTER 1, OR 2, OR 3,..., OR 11 ";
2460    INPUT X
2470    IF X<1 OR X>11 THEN 2440
2480    GOSUB 2770
2490    W=5*B
2500    IF U >= 3*X-2 AND U <= 3*X+3 THEN 2710
2510    GOTO 2590
2515    REM 2520 TO 2550 DEALS WITH BETS ON THE QUATRE
2516    REM PREMIERS AT THE EUROPEAN TABLE AND THE FIRST
2517    REM FIVE AT THE AMERICAN TABLE.
2520    GOSUB 2770
2530    W=(T-1)*6*B+(2-T)*8*B
2540    IF U <= 3 THEN 2710
```

Fig. 2.14 Master program "Roulette" (cont'd)

```
2550    GOTO 2590
2560    PRINT TAB(12)"*** COUP NEUTRE...HERE WE GO AGAIN ***"
2570    P=1
2580    GOTO 550
2590    PRINT LIN(1)
2600    PRINT "YOU JUST LOST "B;M$" !"
2610    PRINT LIN(1)
2620    M=M-B
2630    IF M=0 THEN 3060
2640    PRINT "YOU NOW HAVE "M;M$", DO YOU WANT TO TRY AGAIN ";
2650    INPUT F$
2660    IF F$[1,1]="N" THEN 3140
2670    GOTO 530
2680    IF W>0 THEN 2710
2690    PRINT "YOU BROKE EVEN !"
2700    GOTO 2640
2710    PRINT '7'7'7'7'7'7'7'7'7'7'7'7'7'7
2720    PRINT TAB(4)"*** CONGRATULATIONS ! YOU JUST WON ";
2730    PRINT W;M$" !! ***"
2740    PRINT LIN(1)
2750    M=M+W
2760    GOTO 2640
2770    PRINT LIN(1)
2780    IF P=1 THEN 2820
2790    PRINT TAB(8)"*** LES JEUX SONT FAITS, RIEN NE VA ";
2800    PRINT "PLUS...***"
2810    PRINT LIN(1)
2820    GOTO SGN(U)+2 OF 2880,2850,2830
2830    PRINT TAB(12)"*** "U","A$","B$;K$[37,39+T];C$" ***"
2840    GOTO SGN(B)+1 OF 530,2950
2850    PRINT TAB(25)"*** ZERO ***"
2860    PRINT
2870    GOTO SGN(B)+1 OF 530,2920
2880    PRINT TAB(21)"*** DOUBLE ZERO ***"
2890    PRINT
2900    GOTO SGN(B)+1 OF 530,2910
2910    GOTO G OF 2590,1520,2590,2590,2290,2590,2590,2590
2920    IF H=2 AND G <> 2 THEN 2590
2930    IF P=1 OR (T=2 AND G <> 2 AND G <> 5) THEN 2590
2940    GOTO G OF 2970,1520,2590,2590,2330,2590,2590,2530
2950    PRINT LIN(1)
2960    RETURN
2970    PRINT TAB(10)"*** DO YOU WANT HALF YOUR STAKE BACK ? ***"
2980    PRINT LIN(1)
2990    INPUT F$
3000    IF F$[1,1]="N" THEN 3030
3010    B=B/2
3020    GOTO 2590
3030    P=1
3040    PRINT TAB(13)"*** EN PRISON...HERE WE GO AGAIN ***"
3050    GOTO 550
3060    PRINT "MOTIVATED BY GREED AND CURSED WITH CONGENITAL ";
3070    PRINT "STUPIDITY, YOU"
3080    PRINT "JUST LOST ALL YOUR MONEY. YOU EITHER TAKE THE ";
3090    PRINT "CONSEQUENCES"
3100    PRINT "LIKE A GENTLEMEN, OR YOU WILL HAVE TO WORK OFF ";
3110    PRINT "YOUR BAR-BILL"
3120    PRINT "IN THE KITCHEN !"
3130    GOTO 3160
3140    PRINT LIN(1)
3150    PRINT "GETTING COLD FEET ? AFTER ALL, IT'S ONLY MONEY !"
3160    END
```

Fig. 2.14 Master program "Roulette" (cont'd)

Note how the "table variable T" (introduced in lines 240 and 390) separates the European table from the American table. Note also how the "*à cheval* variable H" (introduced in lines 920 and 950) is used to pass control to the appropriate sections of the program. Finally, there is the ubiquitous variable P, which plays a dual role. Normally 0, it is set to 1 after a zero and it is also set to 1 after a *coup neutre*. It causes control to pass to the "YOU LOST . . ." statement (2600) if the zero comes up a second time in a row (and the player took the "en prison option") and causes the betting procedure and the attendant ballyhoo to be bypassed for the *coup* following the first zero (see line 680). It also causes the betting to be bypassed after a *coup neutre* (see line 2570). These two roles of P can never get mixed up because a *coup neutre* can occur only after an *à cheval* bet on a *chance simple*, and there is no prison for *à cheval* bets! A zero in such a case causes a loss straightaway!

We wish to call particular attention to the computation of the winnings in line 1190 for the case of a *chance simple:*

$$W = B - P * B + (H-1) * B * P = \begin{cases} O \text{ if } P = 1, H = 1 \text{ (zero on last } coup, \text{ not } à \text{ } cheval) \\[1em] B \begin{cases} \text{if } P = O, H = 1 \text{ (no zero, not } à \text{ } cheval) \\ \text{if } P = O, H = 2 \text{ (no zero, } à \text{ } cheval) \\ \text{if } P = 1, \; H = 2 \text{ (} coup \text{ } neutre, \text{ } à \text{ } cheval) \end{cases} \end{cases}$$

Why go through these contortions? Well, if you win when en prison (P = 1), you don't get any money; you break even (see 2680 and 2690). On the other hand, if you win the *coup* following a *coup neutre* (also P = 1), which can only happen on an *à cheval* bet (H = 2), then you do win even money, and the P*B which has been subtracted from B thus has to be restored with + (H−1)*B*P.

Also observe how a bet of 0 dollars (or whatever the currency) in line 720 causes control to be passed from 730 to 2820—bypassing the betting, the chanting, and what have you—and from 2840, 2870, or 2900 back to 530 for the next spin of the wheel. We found this arrangement to be very practical when we want to observe the wheel for a while before embarking on a system.

Next, let us explain the "recognition loops," 790−830 and 1060−1080, which encode the inputs in 780 and 1050, respectively, no matter whether they are in French or English.

We see from

| C | H | A | N | C | E | S |   S | I | M | P | L | E | S |
|---|---|---|---|---|---|---|---|---|---|---|---|---|---|
| E | N |   | P | L | E | I | N |
| C | O | L | O | N | N | E |
| D | O | U | Z | A | I | N | E |

| E | V | E | N |   | C | H | A | N | C | E |
|---|---|---|---|---|---|---|---|---|---|---|
| S | T | R | A | I | G | H | T |   | U | P |
| C | O | L | U | M | N |
| D | O | Z | E | N |

| T | R A | N | S V E R S A L E |
|---|-----|---|---|
| C | A R | R | E |
| S | I X | A | I N |
| Q | U A | T | R E     P R E M I E R S |

| S | T R | E | E T |
|---|-----|---|---|
| C | O R | N | E R |
| S | I X | L | I N E |
| F | I R | S | T      F I V E |

that the first and fourth letters suffice in both cases to distinguish between the different inputs in 780, and we see from

| N | O | I | R |
|---|---|---|---|
| R | O | U | G E |
| P | A | I | R |
| I | M P | A | I R |
| P | A | S | S E |
| M | A N | Q | U E |

| B | L | A | C K |
|---|---|---|---|
| R | E | D | |
| E | V | E | N |
| O | D | D | |
| H | I | G | H |
| L | O | W | |

that the second and third letters will serve that purpose for the input in 1050. E\$ (in 330 and 480) and G\$ (in 1000 and 1040) are defined accordingly.

Next, a word about the strange entry in line 2530,

$$2530 \ W = (T-1)*6*B+(2-T)*8*B$$

In lines 2520 to 2550, we deal simultaneously with a bet on the *quatre premiers* at the European table and the *first five* at the American table. You win in either case if $U \leq 3$ (0, 1, 2, or 3 at the European table and $-1$, 0, 1, 2, 3 at the American table), but the pay-off at the European table is 8 to 1 and at the American table, 6 to 1. By line 2530,

$$W = \begin{cases} 6*B \text{ if and only if } T = 2 \\ 8*B \text{ if and only if } T = 1 \end{cases}$$

and there is the explanation!

Finally, let us comment on the mystifying entry in line 2710. This is the computer's regurgitation of

2710 PRINT "(We depressed the CTRL-key and hit the BELL-key twelve times)"

When control is passed to this line right after a winning *coup*, a bell— or an electronic facsimile thereof—will ring twelve times.

## MODIFICATIONS OF THE COMPUTER PROGRAM

If you decide in favor of the European table on the one hand, or the American table on the other, and want to stick with your choice, you may achieve a substantial simplification of the program displayed in Fig. 2.14 by eliminating everything pertaining to the other one. To eliminate the American table, for example, you omit the following lines: 70–100, 140, 150, 230, 240, 380–520, 540, 560, 570, 870, 890, 900, 970, 1010–1040, 1280–1340, 1410, 1500, 1780, 1800, 1810, 2040–2120, 2180–2210, and 2880–2910. You must also change lines 1830, 1850, 1920, 2530, 2820, and 2930 as follows:

```
1830   H$ = "PMD"
1850   IF F$(1,1) = H$(J,J) THEN 1870
1920   IF I$(1,1) = H$(L,L) THEN 1940
2530   W = 8*B
2820   GOTO SGN(U)+2 OF 2850, 2830
2930   IF P=1 THEN 2590
```

If you want the computer to print out the relative frequencies of, let us say, the various *chances simples* after the print-out of the result of the *coup* (prompted in line 2830), you may proceed as follows:

```
31    DIM F[6]
32    MAT F=ZER[6]
33    N=0
551   N=N+1
552   IF U <= 0 THEN 680
571   N=N+1
572   IF U <= 0 THEN 680
581   F[C2]=F[C2]+1
591   F[C3]=F[C3]+1
621   F[2]=F[2]+1
641   F[1]=F[1]+1
```

```
2831   GOSUB 3131
2851   GOSUB 3131
2881   GOSUB 3131
3131   PRINT "NO.OF PLAYS:REL.FREQU.OF BLACK:RED  :EVEN :ODD  :HIGH
                                                       :LOW  :"
3132   PRINT  USING 3133;N,F[1]/N,F[2]/N,F[3]/N,F[4]/N,F[5]/N,F[6]/N
3133   IMAGE 3X,6D,16X,SD.2D,X,SD.2D,X,SD.2D,X,SD.2D,X,SD.2D,X,SD.2D
3134   RETURN
```

Of greater interest than the relative frequencies, F(J)/N, are the deviations from the norm, 18/37 (or 18/38 = 9/19 at the American table). You may therefore wish to modify this latest addition as follows:

```
671    FOR J=1 TO 6
672    D[J]=(T-1)*((F[J]/N)-(9/19))-(T-2)*((F[J]/N)-(18/37))
673    NEXT J
3131   PRINT "NO.OF PLAYS:DEVIATIONS  :BLACK:RED  :EVEN :ODD  :HIGH
                                                       :LOW  :"
3132   PRINT  USING 3133;N,D[1],D[2],D[3],D[4],D[5],D[6]
```

```
34    DIM D[6]
552   IF U <= 0 THEN 671
572   IF U <= 0 THEN 671
```

We have superimposed the latter on our program in Fig. 2.14 and tried it out. Here is a sample run:

```
DO YOU WANT TO PLAY AT THE EUROPEAN TABLE OR AT THE
AMERICAN TABLE ?AMERICAN


HOW MUCH MONEY DO YOU HAVE ?10000


                    *** GOOD LUCK ***



                *** HOW MUCH DO YOU WANT TO BET ? ***

?0
                *** 35   ,BLACK ,ODD     AND HIGH    ***
NO.OF PLAYS:DEVIATIONS   :BLACK:RED  :EVEN :ODD  :HIGH :LOW  :
        1                +0.53 -0.47 -0.47 +0.53 +0.53 -0.47


                *** HOW MUCH DO YOU WANT TO BET ? ***

?0
                *** 29   ,BLACK ,ODD     AND HIGH    ***
NO.OF PLAYS:DEVIATIONS   :BLACK:RED  :EVEN :ODD  :HIGH :LOW  :
        2                +0.53 -0.47 -0.47 +0.53 +0.53 -0.47


                *** HOW MUCH DO YOU WANT TO BET ? ***

?0
                *** 15   ,BLACK ,ODD     AND LOW     ***
NO.OF PLAYS:DEVIATIONS   :BLACK:RED  :EVEN :ODD  :HIGH :LOW  :
        3                +0.53 -0.47 -0.47 +0.53 +0.19 -0.14



                *** HOW MUCH DO YOU WANT TO BET ? ***

?0
                    *** DOUBLE ZERO ***
NO.OF PLAYS:DEVIATIONS   :BLACK:RED  :EVEN :ODD  :HIGH :LOW  :
        4                +0.28 -0.47 -0.47 +0.28 +0.03 -0.22



                *** HOW MUCH DO YOU WANT TO BET ? ***

?0
                *** 9    ,RED   ,ODD     AND LOW     ***
NO.OF PLAYS:DEVIATIONS   :BLACK:RED  :EVEN :ODD  :HIGH :LOW  :
        5                +0.13 -0.27 -0.47 +0.33 -0.07 -0.07


                *** HOW MUCH DO YOU WANT TO BET ? ***

?0
```

```
          ***   25    ,RED    ,ODD    AND HIGH   ***
NO.OF PLAYS:DEVIATIONS   :BLACK:RED  :EVEN :ODD  :HIGH :LOW  :
       6                  +0.03 -0.14 -0.47 +0.36 +0.03 -0.14


          *** HOW MUCH DO YOU WANT TO BET ? ***

?2000


          *** MESSIEURS, FAITES VOS JEUX ...***


?EVEN CHANCE
ENTER 'RED', OR 'BLACK', OR 'EVEN', OR 'ODD',
OR 'LOW', OR 'HIGH' ?EVEN


       *** LES JEUX SONT FAITS, RIEN NE VA PLUS...***


          *** 9     ,RED    ,ODD    AND LOW    ***
NO.OF PLAYS:DEVIATIONS   :BLACK:RED  :EVEN :ODD  :HIGH :LOW  :
       7                  -0.05 -0.05 -0.47 +0.38 -0.05 -0.05



YOU JUST LOST   2000    DOLLARS !

YOU NOW HAVE   8000    DOLLARS. DO YOU WANT TO TRY AGAIN ?NO

GETTING COLD FEET ? AFTER ALL, IT'S ONLY MONEY !
DONE
```

It didn't seem to help much, or did it?

The same approach may be taken—with proper modifications, of course—to obtain the relative frequencies with which certain numbers appear, or certain columns, certain dozens, and so forth.

The reader may wish to modify our program further to permit prisons of higher degree. (After a zero, let $P = P+1$, and after a subsequent win, let $P = P-1$. The stake is let out of prison if $P = 0$; it is placed in double prison if $P = 2$, etc.) He may also incorporate the house minimum and house maximum into the program so that the computer will reject bets below the minimum and above the maximum.

So far, our program does not permit us to bet on any combinations other than the ones already discussed. This deficiency is easily remedied. We will illustrate a modification of our program that accommodates a progression on the so-called *action numbers*. In principle, action numbers are seven numbers that are fairly evenly distributed around the wheel, such as 10, 11, 12, 13, 14, 15, and 33 on the American wheel. Because of their even distribution, the ball will have to bounce around one of them eventually before it settles down in one of the compartments (hence the term *action numbers*, since the action takes place near one of them). The reader may select his own action numbers. On the European wheel, 4, 5, 6, 7, 8, 9, and 26 would be a reasonable choice.

Here is our strategy: We lay $1 (*en plein*—straight up) on each of these seven numbers. (To bet on a set of numbers such as the ones mentioned, one simply puts six chips on the *sixain* 10–11–12–13–14–15 or 4–5–6–7–8–9, respectively, and the seventh chip on 33 or 26, respectively.) If we lose, we are $7 in the hole and we quit. If we win, we make a net gain of $35 − $6, or $29, and we triple our bet ($3 per number for a total of $21). If we lose now, we still hold $8 ($29 − $21), but if we win, we have a total of $116 ($29 + $105 − $18). Here is how we proceed: If we lose, we reduce our bet on the next *coup* to one-third of what it was (on the losing *coup*), provided that the reduced bet amounts to at least $1 per number. If not, we get out. If we win, however, we again triple our bet until we either get dizzy or the house limit is attained. (The house maximum on an *en plein* bet in the Monte Carlo "kitchen" is 300 francs, and in the *salles privées*, 600 francs. In Atlantic City, the house maximum on a single number is $100.) If we keep on winning, we just keep tripling our bet or betting the house maximum, whichever is lower. Now, unless we lose the first *coup*, or win the first *coup* but lose the second and third *coup*—the one or the other will happen with a probability of 0.93838— and if we play strictly by our rule, we will have made a net gain of at least $16 and possibly as much as $1160 after the fourth *coup*.

How do we accommodate such a play in our program? Well, we'll simply incorporate it as an option under the *straight up* play at the American table, as follows:

```
1280   IF H=2 THEN 1290
1281   PRINT "DO YOU WANT TO BACK THE NUMBERS 10,11,12,13,14,15,33 ";
1282   INPUT F$
1283   IF F$[1,1]="N" THEN 1290
1284   G=3
1285   GOSUB 2770
1286   W=29*B/7
1287   IF (U<10 OR U>15) AND U <> 33 THEN 2600
1288   GOTO 2710
```

Notice that we set $G = 3$ in line 1284 to insure a return of the control to the "YOU LOST . . ." message in line 2600 after a *zero* or a *double zero*.

We tried the system on our computer, and here is what happened:

```
DO YOU WANT TO PLAY AT THE EUROPEAN TABLE OR THE
AMERICAN TABLE ?AMERICAN
```

HOW MUCH MONEY DO YOU HAVE ?70


                        *** GOOD LUCK ***



                *** HOW MUCH DO YOU WANT TO BET ? ***
?70


                *** MESSIEURS, FAITES VOS JEUX ...***

?STRAIGHT UP
SPLIT ? (ENTER YES OR NO) ?NO
DO YOU WANT TO BACK THE NUMBERS 10,11,12,13,14,15,33 ?YES

            *** LES JEUX SONT FAITS, RIEN NE VA PLUS...***


            *** 11   ,BLACK ,ODD   AND LOW    ***


        *** CONGRATULATIONS ! YOU JUST WON  290  DOLLARS !! ***


YOU NOW HAVE  360  DOLLARS. DO YOU WANT TO TRY AGAIN ?YES


                *** HOW MUCH DO YOU WANT TO BET ? ***
?210


                *** MESSIEURS, FAITES VOS JEUX ...***


?STRAIGHT UP
SPLIT ? (ENTER YES OR NO) ?NO
DO YOU WANT TO BACK THE NUMBERS 10,11,12,13,14,15,33 ?YES


            *** LES JEUX SONT FAITS, RIEN NE VA PLUS...***


            *** 34   ,RED   ,EVEN  AND HIGH   ***


YOU JUST LOST  210  DOLLARS !

YOU NOW HAVE  150  DOLLARS. DO YOU WANT TO TRY AGAIN ?YES


                *** HOW MUCH DO YOU WANT TO BET ? ***
?70

*** MESSIEURS, FAITES VOS JEUX ...***

?STRAIGHT UP
SPLIT ? (ENTER YES OR NO) ?NO
DO YOU WANT TO BACK THE NUMBERS 10,11,12,13,14,15,33 ?YES

      *** LES JEUX SONT FAITS, RIEN NE VA PLUS...***


      ***  2     ,BLACK ,EVEN   AND LOW    ***


YOU JUST LOST  70    DOLLARS !


YOU NOW HAVE  80    DOLLARS. DO YOU WANT TO TRY AGAIN ?NO


GETTING COLD FEET ? AFTER ALL, IT'S ONLY MONEY !

DONE


Well, it didn't work out the way we hoped it would!

One might program the computer to play a system all by itself and print out only the final result. For reasons of simplicity, let us assume that we want the computer to play a martingale at the American table, backing EVEN, with 1 dollar being the house minimum and 1000 dollars, the house maximum. The program is as follows:

```
10   A=0
20   B=1
30   N=0
40   U=INT(RND(1)*38)-1
50   N=N+1
60   IF U>0 AND U/2=INT(U/2) THEN 140
70   A=A-B
80   B=2*B
90   IF B <= 1000 THEN 40
100  PRINT "THE NEXT BET IN THE MARTINGALE PROGRESSION WOULD"
110  PRINT "EXCEED THE HOUSE MAXIMUM. YOU ARE "-A" DOLLARS"
120  PRINT "IN THE HOLE AFTER "N" COUPS !"
130  GOTO 150
140  PRINT "YOU WON 1 DOLLAR AFTER "N" COUPS."
150  END
```

We ran this program several times, and here is what happened:

```
          RUN

          YOU WON 1 DOLLAR AFTER  4      COUPS.

          DONE
```

```
RUN

YOU WON 1 DOLLAR AFTER   3      COUPS.

DONE

RUN

YOU WON 1 DOLLAR AFTER   1      COUPS.

DONE

RUN

YOU WON 1 DOLLAR AFTER   8      COUPS.

DONE

RUN


THE NEXT BET IN THE MARTINGALE PROGRESSION WOULD
EXCEED THE HOUSE MAXIMUM. YOU ARE   1023      DOLLARS
IN THE HOLE AFTER   10     COUPS !

DONE
```

Doesn't that scare you?

Finally, let us mention how one may introduce some extra excitement by having the computer print out a sequence of numbers before settling on the winning number, U. An extra thrill is achieved by introducing a variable time-delay between the print-out of two such consecutive numbers. Here is one way of modifying our program in Fig. 2.14 to do just this at the European table:

```
 61   DIM W[74]
550   FOR J=1 TO 37
551   READ N
552   W[J]=N
553   NEXT J
554   DATA 0,32,15,19,4,21,2,25,17,34,6,27,13,36,11,30,8,23
555   DATA 10,5,25,16,33,1,20,14,31,9,22,18,29,7,28,12,35,3,26
```

(We have taken the option of printing the numbers in the sequence as they appear on the wheel. If you wish to simulate the erratic bouncing of the ball, you will have to change the program accordingly.)

```
556   RESTORE
557   FOR J=38 TO 74
558   READ N
559   W[J]=N
560   NEXT J
```

```
561   RESTORE
562   S=INT(RND(1)*37)+1
563   V=INT(RND(1)*37)+38
564   U=W[V]
565   GOTO 580


2811  FOR J=S TO V
2812  PRINT W[J];
2813  FOR I=1 TO 16*J
2814  X=1
2815  NEXT I
2816  NEXT J
2817  PRINT
```

Observe how we induced the variable time-delay between the print-out of two numbers by means of "nonsense loop" 2813–2815, which produces no visible result but keeps the computer "idling" for longer periods of time.[1] What we put into line 2814 does not really matter so long as it is acceptable to the computer and does not foul up other parts of our program. It is best to use some irrelevant assignment statement.

We leave it to the reader and his imagination and skill to conceive and execute other possible modifications, amplifications, and simplifications of our computer program. Suffice it here to say that the number of possibilities is staggering.

---

[1] Because of the peculiar way in which computers do their arithmetic, the time delay does not increase monotonically as one would normally expect, but somewhat erratically.

# Chapter 3

# chemin-de-fer

```
*****************
* CHEMIN-DE-FER *
*****************


  *** STAND BY - THE CARDS ARE BEING SHUFFLED ***


 *** PLEASE CUT - BY ENTERING A NUMBER BETWEEN 1 AND 312 ***


?69
PLEASE IDENTIFY YOURSELF ?HANS SAGAN


BON SOIR, MONSIEUR SAGAN, COMMENT ALLEZ-VOUS ? THE OTHER PLAYERS
ARE SIR HILARY BRAY,LE MONSTRE DE LILLE,THE COMTE DE BLEUCHAMP,
COMMANDER BOND, AND SIR PALMER EPSOM-OAKS. YOU WILL, IF YOU
PLEASE, PLAY IN THE FOLLOWING ORDER:


             1     SIR PALMER
             2     M. SAGAN
             3     SIR HILARY
             4     LE MONSTRE
             5     M. LE COMTE
             6     CMDR. BOND


SIR PALMER , HOW MUCH MONEY DO YOU WISH TO PUT UP ? 5000
M. SAGAN    , HOW MUCH MONEY DO YOU WISH TO PUT UP ?4500
ONLY MULTIPLES OF THOUSAND FRANCS, PLEASE !
M. SAGAN    , HOW MUCH MONEY DO YOU WISH TO PUT UP ?5000
SIR HILARY , HOW MUCH MONEY DO YOU WISH TO PUT UP ? 3000
LE MONSTRE , HOW MUCH MONEY DO YOU WISH TO PUT UP ? 9000
M. LE COMTE, HOW MUCH MONEY DO YOU WISH TO PUT UP ? 7000
CMDR. BOND , HOW MUCH MONEY DO YOU WISH TO PUT UP ? 10000


CMDR. BOND  HAS THE BANK FOR  10000     FRANCS.
```

*Fig. 3.1 Run of the program "Chemin-de-Fer"*

*** BANCO DE 10.0 MILLE ***


*** MESSIEURS, FAITES VOS JEUX... ***


SIR PALMER , ENTER YOUR BET ! 300
M. SAGAN   , ENTER YOUR BET !?6000
YOU DON'T HAVE ENOUGH MONEY. ENTER AN AMOUNT
NOT TO EXCEED  5000      ?2000
SIR HILARY , ENTER YOUR BET ! 1300
LE MONSTRE , ENTER YOUR BET ! 1200
M. LE COMTE, ENTER YOUR BET ! 1000


        ** LES JEUX SONT FAITS, RIEN NE VA PLUS **


                                    *** FIVE   OF HEARTS   ***
                                    *** QUEEN OF SPADES   ***

DO YOU WANT ANOTHER CARD ?YES

HERE IS THE                         *** TWO    OF DIAMONDS ***

THE BANKER DRAWS THE

*** SEVEN OF DIAMONDS ***

TO THE

*** FOUR   OF CLUBS    ***
*** SEVEN OF HEARTS   ***


FOR A COUNT OF  8     TO THE OPPOSITION'S COUNT OF  7     TO
WIN 5800     FRANCS. 290  FRANCS GO INTO THE CAGNOTTE
AND 10   FRANCS ARE 'UN POURBOIRE' FOR THE CHEF DE JEU.


THE PLAYERS ARE NOW HOLDING THE FOLLOWING AMOUNTS:

                    SIR PALMER   ...  4700
                    M. SAGAN     ...  3000
                    SIR HILARY   ...  1700
                    LE MONSTRE   ...  7800
                    M. LE COMTE  ...  6000
                    CMDR. BOND   ...  15500


CMDR. BOND  HAS THE BANK FOR  15500     FRANCS.


            *** BANCO DE 15.5 MILLE ***


        *** MESSIEURS, FAITES VOS JEUX... ***


SIR PALMER , ENTER YOUR BET ! 700
M. SAGAN   , ENTER YOUR BET !?500


*Fig. 3.1 Run of the program "Chemin-de-Fer" (cont'd)*

```
SIR HILARY , ENTER YOUR BET ! 800
LE MONSTRE , ENTER YOUR BET ! 200
M. LE COMTE, ENTER YOUR BET ! 600


          ** LES JEUX SONT FAITS, RIEN NE VA PLUS **


                                *** THREE OF  CLUBS      ***
                                *** TEN    OF  CLUBS      ***

DO YOU WANT ANOTHER CARD ?YES

HERE IS THE                     *** FIVE   OF DIAMONDS ***


THE BANKER STAYS WITH

*** ACE    OF SPADES    ***
*** SIX    OF DIAMONDS ***


FOR A COUNT OF  7      TO THE OPPOSITION'S COUNT OF   8       TO
LOSE  2800     FRANCS. THE BANK PASSES TO THE NEXT PLAYER.

THE PLAYERS ARE NOW HOLDING THE FOLLOWING AMOUNTS:


                    SIR PALMER   ...   5400
                    M. SAGAN     ...   3500
                    SIR HILARY   ...   2500
                    LE MONSTRE   ...   8000
                    M. LE COMTE ...   6600
                    CMDR. BOND   ...  12700


SIR PALMER  HAS THE BANK FOR  5400      FRANCS.


          *** BANCO DE  5.4 MILLE ***


          *** MESSIEURS, FAITES VOS JEUX... ***


M. SAGAN    , ENTER YOUR BET !?1500
SIR HILARY , ENTER YOUR BET ! 400
LE MONSTRE , ENTER YOUR BET ! 800
M. LE COMTE, ENTER YOUR BET ! 800
CMDR. BOND , ENTER YOUR BET ! 700


          ** LES JEUX SONT FAITS, RIEN NE VA PLUS **


                                *** THREE OF SPADES      ***
                                *** FOUR  OF DIAMONDS ***
DO YOU WANT ANOTHER CARD ?NO

THE BANKER STAYS WITH
```

*Fig. 3.1 Run of the program "Chemin-de-Fer" (cont'd)*

```
*** SEVEN OF CLUBS      ***
*** KING  OF SPADES     ***



                    *** COUP NEUTRE ***


                              *** KING   OF DIAMONDS ***
                              *** TEN    OF CLUBS     ***

DO YOU WANT ANOTHER CARD ?YES

HERE IS THE                   *** FOUR  OF SPADES     ***

THE BANKER DRAWS THE

*** EIGHT OF DIAMONDS ***

TO THE

*** ACE   OF CLUBS      ***
*** QUEEN OF DIAMONDS ***


FOR A COUNT OF  9     TO THE OPPOSITION'S COUNT OF   4      TO
WIN 4200    FRANCS. 210  FRANCS GO INTO THE CAGNOTTE
AND 90   FRANCS ARE 'UN POURBOIRE' FOR THE CHEF DE JEU.


THE PLAYERS ARE NOW HOLDING THE FOLLOWING AMOUNTS:



                    SIR PALMER   ...  9300
                    M. SAGAN     ...  2000
                    SIR HILARY   ...  2100
                    LE MONSTRE   ...  7200
                    M. LE COMTE  ...  5800
                    CMDR. BOND   ...  12000


SIR PALMER  HAS THE BANK FOR  9300       FRANCS.

                    *** BANCO DE  9.3 MILLE ***


              *** MESSIEURS, FAITES VOS JEUX... ***


M. SAGAN    , ENTER YOUR BET !?500
SIR HILARY , ENTER YOUR BET ! 400
LE MONSTRE , ENTER YOUR BET ! 600
M. LE COMTE, ENTER YOUR BET ! 1300
CMDR. BOND , ENTER YOUR BET ! 1300


           ** LES JEUX SONT FAITS, RIEN NE VA PLUS **
```

*Fig. 3.1 Run of the program "Chemin-de-Fer" (cont'd)*

```
                                    *** TWO   OF SPADES   ***
                                    *** QUEEN OF CLUBS    ***
```

DO YOU WANT ANOTHER CARD ?YES

```
HERE IS THE                         *** SIX    OF HEARTS   ***
```

THE BANKER STAYS WITH

```
*** THREE OF DIAMONDS ***
*** FOUR  OF CLUBS    ***
```

FOR A COUNT OF  7      TO THE OPPOSITION'S COUNT OF  8      TO
LOSE  4100      FRANCS. THE BANK PASSES TO THE NEXT PLAYER.


THE PLAYERS ARE NOW HOLDING THE FOLLOWING AMOUNTS:

```
                    SIR PALMER   ...   5200
                    M. SAGAN     ...   2500
                    SIR HILARY   ...   2500
                    LE MONSTRE   ...   7800
                    M. LE COMTE  ...   7100
                    CMDR. BOND   ...   13300
```

M. SAGAN    HAS THE BANK FOR  2500      FRANCS.


```
              *** BANCO DE  2.5 MILLE ***


          *** MESSIEURS, FAITES VOS JEUX... ***
```

SIR PALMER , ENTER YOUR BET ! 1200
SIR HILARY , ENTER YOUR BET ! 100
LE MONSTRE , ENTER YOUR BET ! 400
M. LE COMTE, ENTER YOUR BET ! 500
CMDR. BOND , ENTER YOUR BET !

```
          *** BANCO - LE BANCO EST FAIT ***
```

CMDR. BOND  BETS  2500      FRANCS ! ALL OTHER BETS ARE VOID.


```
       ** LES JEUX SONT FAITS, RIEN NE VA PLUS **
```

```
*** EIGHT OF DIAMONDS ***
*** FIVE  OF SPADES   ***
```

```
              *** LE GRAND ***
```

CMDR. BOND  IS SITTING ON

```
                                    *** FOUR  OF DIAMONDS ***
                                    *** FIVE  OF DIAMONDS ***
```

*Fig. 3.1 Run of the program "Chemin-de-Fer" (cont'd)*

```
FOR A TOTAL COUNT OF   9    , WHILE YOUR COUNT IS   3    AND YOU
LOSE   2500     FRANCS. THE BANK PASSES TO THE NEXT PLAYER.

THE PLAYERS ARE NOW HOLDING THE FOLLOWING AMOUNTS:

                    SIR PALMER   ...   5200
                    M. SAGAN     ...   0
                    SIR HILARY   ...   2500
                    LE MONSTRE   ...   7800
                    M. LE COMTE  ...   7100
                    CMDR. BOND   ...   15800


YOU WERE TAKEN TO THE CLEANERS BY EXPERTS.

THE CASINO MADE 500  FRANCS IN COMMISSIONS AND THE
CHEF DE JEU MADE 100  FRANCS IN TIPS.

DONE
```

Fig. 3.1 Run of the program "Chemin-de-Fer" (cont'd)

We hope that Ian Fleming, wherever he may be, won't mind our borrowing some of his fictional characters (Ref. [4], p. 22 ff).

## HOW THE GAME IS PLAYED

The game chemin-de-fer has very little in common with the games that have been discussed hitherto, other than being a game of chance. While the entertainment level of trente-et-quarante and roulette without betting is about the same as that of watching paint dry, chemin-de-fer would have some limited entertainment value on it's own merit even if no money changed hands. Strategies come into play, and there is even room for playing hunches—provided that the house rules are not too strict and leave the players some flexibility.

Also, chemin-de-fer—in contrast to trente-et-quarante and roulette—pits player against player and not player against some monolithic consortium running a casino. Here, the casino merely plays the role of the landlord, supplying the equipment (table with money box—cagnotte—and discard cylinder, a shoe—sabot—from which the cards are dealt, and a palette to slide money and cards hither and yon) and the personnel (a chef de jeu, or pit boss, and a croupier, or houseman) in exchange for a fixed percentage (usually 5 percent) of the banker's take.

Finally, chemin-de-fer is a game for high rollers. It is not unusual to see players casually handle stacks of century notes and occasional grands! Some "rug joints" (fashionable casinos) require players to be formally attired. So, if you are wearing your kudzu-alliance T-shirt and cut-off jeans, we suggest that you get yourself a dollars worth of nickels from the change window and seek your thrills and fortunes at the slot machines.

Chemin-de-fer, or "Shimmy" or "Chemmy" (as it is called by many monoglot players in this country), is a card game. Two hands of two cards each are dealt down—one for the players and one for the dealer-banker. Each party has the option of taking a third card, which is to be dealt up. The one who comes closest to a count of 9 wins. We are now going to explain the game in detail.

The game is played on a kidney-shaped table and requires, in principle, at least two players, with no upper limit—again in principle—on the number of players. In the real world, there usually are lower and upper limits imposed on the number of participants (many casinos require nine or twelve players). The players arrange themselves in a certain order (which they adhere to for the duration of the game) and bid for the bank. The highest bidder gets to be banker and stays banker as long as he keeps winning. When he loses (misses a pass), the bank passes to the player on his right, who may refuse it and pass it on, in turn, in the counterclockwise direction, until it is accepted. The game is called *chemin-de-fer* (railroad) because the bank travels around the table just like a model railroad train. Every time the banker wins, the house rakes in 5 percent of the winnings. This "rent money" is placed into a *cagnotte* (money box).

Once a player has established himself as the banker and put his money on the line (the amount which he "bid" to obtain the bank), six or eight decks of ordinary playing cards are shuffled—first one deck at a time, then all together—cut, and the first three cards are discarded ("burned").

Now comes the betting. The players place their (even-money) bets in their prearranged order, starting with the player on the banker's right and proceeding in counterclockwise direction. There is no point for anyone to bet an amount that is higher than the amount left in the bank after the preceding bets have been subtracted, because the banker is not responsible for any pay-offs that exceed the amount he has put up as the bank. There is one exception. Any player has the right to go

★★★banco★★★

that is, bet the entire amount that is in the bank ("fade the bank") even if other bets have already been placed. Going *banco* voids all other bets, and the game is now between the banker and the player who has gone *banco*. If more than one player want to go *banco*, the one first in line has precedence. The banker may refuse this bet and pass the bank to the nearest player in line who is willing to take it. If a player loses his *banco* bet, he has precedence over all other players for going *banco* on the next turn (*banco suivi*). If he does not go *banco* and nobody else does either, he may go *banco avec la table* (with the table), which means that the others may bet whatever they please but that he is prepared to cover the difference that is required to fade the bank. *Warning:* If you bet an amount in excess of what you can cover and you lose (*le coup du déshonneur*), the casino may make up the difference (to keep the other players from doing you bodily harm) but you will be blackballed from here to Nishnij-Novgorod!

Two hands of cards are dealt when the betting is over—one hand for the players, who play this hand collectively through one representative (the "active player"—usually the one who placed the largest bet), and one for the banker, as follows: The banker deals one card down to the player, one card down to himself, and then repeats this deal once more. Each party now has two cards down and takes a careful peek at them. It is the objective of the game to come as close as possible to a count of 9. Court cards (face cards) count 10, and the other cards count their face value, *but* the count is taken modulo 10, meaning that only the last digit of the count counts! For example, if you have 26, then your count is 6, and if you have 19, then your count is 9, and so on. (In this reckoning, court cards and the *ten* really count 0 and the highest possible count is 9.)

If either the player or the banker (or both) are sitting on a count of 8 or 9, they flip their cards up and shout

$$\star\star\star le\ petit \star\star\star$$

(the little one) if the count is 8, or

$$\star\star\star le\ grand \star\star\star$$

(the big one) if the count is 9, and the game is over. The one with the higher count wins. If both have the same count, then there is a

$$\star\star\star coup\ neutre \star\star\star$$

(undecided turn), bets are held over, and the deal is repeated.

If there is no "natural" (*le petit* or *le grand*), the banker asks the player if he wants another card. If he does, he says "*un card*" or "hit me," and the card is dealt face up. Then the banker cogitates whether or not he wants another card. If he does, he also takes it face up. That's it. No more cards. The cards are turned over, and, unless there is a *coup neutre* (both with the same count), the banker or the player wins.

If the banker wins, the *croupier* rakes in all the stakes, puts 5 percent into the *cagnotte*, and pushes the rest towards the banker. The banker is not required to put up more than what he contracted for when he acquired the bank. He may put his winnings *en garage* (actually, what's left of his winnings after the house has been paid and the *chef de jeu* has received a generous tip—*pourboire pour le personnel*), that is, remove them from the stake with the original amount remaining in the bank.

If the banker loses, each player is paid off even money from the bank, and the bank is passed to the next player in line, who may refuse it and pass it on, in turn.

Note: In the European version of baccarat, of which chemin-de-fer is a variant, the bank is run by a casino or syndicate (Ref. [3], p. 161), and three hands are dealt—one for the banker and two for the players. Players may bet on one or the other or (*à cheval*) on both.

The American version of baccarat is much closer to chemin-de-fer. Only two hands are dealt. One is called the *dealer's hand* and the other the *player's*. Players may bet on either hand. Strict rules govern the draws to the dealer's hand as well as the player's hand. The player who wins a bet on the dealer's hand has to pay a certain percentage of his winnings to the house.

### WHEN TO STAY AND WHEN TO DRAW

There is a fundamental difference between the position of the active player (the one who plays the hand for all players in opposition to the banker) and the banker. Since it is the player who must decide first whether he wants to be hit or not, he has no clue whatever as to what the banker is sitting on. He shouldn't rely on a banker's facial expression. There are those who manage to smile smugly while sitting on a count of zero. (To do so is *not* easy when you have $1000 riding on your hand!) The active player has to make his decision solely on the basis of what he is holding. The banker, by contrast, has some indication of the player's position: first, from the player's decision either to stay or to draw, and, second, if the player does draw, from the extra card that is dealt face up.

Since the active player plays the hand for all the other players as well, strict rules are imposed on his strategy to avoid embarrassment, recriminations, and possible altercations. The player *must* draw with a count of 4 or less, and he *must* stay with a count of 6 or 7. (Remember that if his count is 8 or 9, the question never arises because *le petit* or *le grand* terminates the coup.) He has the choice to stay or draw with a count of 5. (In baccarat, however, he *must* draw on 5.)

The banker is free to make his own decisions. In some cases, it is obvious what this decision must be. Suppose, for example, that the banker's count is 3 and the active player draws a 2. This means that the player now has a count of 2, 3, 4, 5, 6, or 7 if he draws on 5 or a count of 2, 3, 4, 5, or 6 if he does not. There is no question but that the banker has to draw. An ace, 2, 3, 4, 5, or 6 will improve his hand, while a 7, 8, or 9 will make it worse (a 10, jack, queen, or king won't change it at all). Let us consider another example. Suppose that the bank has a count of 7. No matter whether the player draws on 5 or stays on 5, the banker will stay because only a 1 or a 2 could improve his hand while all other cards would make it worse or not change it.

Not all cases are that easy to decide. The application of some fundamental principles of *game theory*, however, lead to the *optimal mixed strategy* of Table 3.1 (Ref. [2], p. 214).

### Table 3.1 Optimal mixed strategy

*(a) Strategy for Player*

| If his count is | The player |
|---|---|
| 0, 1, 2, 3, 4 | draws |
| 5 | draws with probability 9/11 and stays with probability 2/11 |
| 6, 7 | stays |

*(b) Strategy for Banker*

| If his count is | The banker | |
|---|---|---|
| 0, 1, 2 | | draws |
| 3 | stays if player draws an 8 | draws in all other cases |
| 4 | stays if player draws 1, 8, 9, 10, J, Q, K | draws if player stays or draws 2, 3, 4, 5, 6, 7 |
| 5 | stays if player draws 1, 2, 3, 8, 9, 10, J, Q, K | draws if player stays or draws 4, 5, 6, 7 |
| 6 | stays if player draws 1, 2, 3, 4, 5, 8, 9, 10, J, Q, K and with probability 1429/2288 if player stays | draws if player draws 6, 7, and with probability 859/2288 if player stays |
| 7 | stays | |

Many modern casinos do not allow the banker an option if he has a count of 6 and the player stays. He is required to stay in that case. This, incidentally, is an optimal mixed strategy if the player stays or draws on a count of 5 with probability ½ each (Ref. [2], p. 215). In baccarat, the banker must also stay with a count of 6 if the player stays.

Assuming that, in the long run, every one of N+1 players is a player N times as often as a banker, that all players bet the same amount, and that the casino rakes in 5 percent of all the banker's winnings, one can show that the *expected loss* is somewhere between $11 and $12 per $1,000 wager, depending on the house rules and the strategy (Ref. [2], p. 215; Ref. [14], p. 205). This expectation is better than that at trente-et-quarante or roulette.

Side bets used to be offered to the player at a pay-off of 9 to 1 that the banker's down cards represent a count of 9, of 9 to 1 that they represent a count of 8, and of 4 to 1 that they represent a count of 8 or 9. The expected loss for all these bets is $53.25 per $1,000 wager. It appears that Edward Thorp (Ref. [13])—who invented the counting system for blackjack and then successfully applied the same principle (to casinos' humiliation) to betting on natural 8s and 9s—is responsible for the removal of this betting option (Ref. [7], p. 99).

Since the banker has a slight advantage over the player, it is considered important to make your pile while you are the banker. The superstitious player believes that if he can win as banker three times in a row, then he is off to the races! There is no basis in fact for this assumption. James Bond, in a precious moment of lucidity, recognizes that "cards have no memory" (Ref. [4], p. 27).

### THE COMPUTER PROGRAM

The core of the computer program that produced the print-out in Fig. 3.1 is the shuffling and dealing of the cards, the counting, and the decision when to stay and when to draw. Everything else is window dressing. To explain how we handled these four activities, we find it most convenient to develop a simple program that does just that.

We shuffle six decks of cards just as we did in Chap. 1 (see p.14) in lines 20 to 300 with *one* difference: Since the count is modulo 10, we have to compute the value G(J) of the Jth card as follows. We replace the old G in line 290 by O:

$$290 \quad O(J) = V(J) \text{ MIN } 10$$

and compute G in terms of O:

$$300 \quad G(J) = O(J) - 10*INT(O(J)/10)$$
$$310 \quad \text{NEXT J}$$

If eight decks are desired, the program has to be modified as explained on p.15. Below, we reproduce, for the convenience of the reader, the card shuffling part of our program with the above modification. We start the line numbers with 30 to leave room for two DIM statements that are to be added later.

```
30   PRINT "STAND BY - THE CARDS ARE BEING SHUFFLED."
40   FOR W=0 TO 5
50   FOR J=1+52*W TO 52+52*W
60   S[J]=J
```

```
70    NEXT J
80    FOR I=52+52*W TO 2+52*W STEP -1
90    J=INT(RND(1)*I)+1
100   T=S[J]
110   S[J]=S[I]
120   S[I]=T
130   NEXT I
140   NEXT W
150   FOR I=312 TO 2 STEP -1
160   J=INT(RND(1)*I)+1
170   T=S[J]
180   S[J]=S[I]
190   S[I]=T
200   NEXT I
210   PRINT "PLEASE CUT - BY ENTERING A NUMBER BETWEEN 1 AND 312."
220   INPUT Z
230   FOR J=1 TO 312
240   T[J]=S[J]+Z-312*INT((S[J]+Z-1)/312)
250   NEXT J
260   FOR J=1 TO 312
270   S[J]=T[J]-52*INT((T[J]-1)/52)
280   V[J]=S[J]-13*INT((S[J]-1)/13)
290   L[J]=INT((S[J]-1)/13)
300   O[J]=V[J] MIN 10
310   G[J]=O[J]-10*INT(O[J]/10)
320   NEXT J
330   P=0
```

For the computer to spell out the names of the cards, when dealing, we need

```
340   V$[1,45]="ACE  TWO   THREEFOUR FIVE SIX   SEVENEIGHTNINE "
350   V$[46,65]="TEN   JACK QUEENKING "
360   S$="CLUBS    DIAMONDSHEARTS  SPADES  "
```

(Note: In chemin-de-fer, the suit is quite immaterial. Still, it adds realism and class to our simulation if we list the suit along with the value of the card.)

No matter whether the computer acts as banker or player, the last up card to be dealt to the banker in a coup is either the fifth or the sixth card that is dealt in that particular coup, depending on whether or not the active player asked for a third card. If he does, three cards have been revealed to the operator at this point, and if he does not, then only two have been revealed. We use the variable N to keep count of these cards and set it at zero at the beginning of every coup. We use P to keep track of the total number of cards that have been played.

```
370   N=0
380   P=P+N
390   N=0
```

We advance N by one unit every time the name of a card is printed out by the computer. At the end of a coup, N cards have been used, and P (which is set to zero, after the shuffle, in line 330) is increased by N units in line 380.

We give the operator the choice to be either the active player or the banker:

```
400   PRINT "DO YOU WANT TO BE BANKER OR PLAYER ";
410   INPUT A$
420   IF A$[1,1]="B" THEN 740
430   FOR J=P+1 TO P+3 STEP 2
440   GOSUB 990
450   NEXT J
```

We put the instruction to print out the specified card(s) in a subroutine because, before we are through, we will have to ask for such a print-out at four different points in our program:

```
990    PRINT V$[1+5*(V[J]-1),5+5*(V[J]-1)]" OF "S$[1+8*L[J],8+8*L[J]]
1000   N=N+1
1010   RETURN
```

Note that the active player gets the first card and the third card (the second and fourth going to the banker), and recall what was said about N.

```
460   PRINT "DO YOU WANT ANOTHER CARD ";
470   INPUT A$
480   IF A$[1,1]="N" THEN 540
490   PRINT "HERE IS THE"
500   J=P+5
510   GOSUB 990
520   Z=G[P+5]
530   GOTO 550
540   Z=-1
```

For the computer as banker to decide on whether to draw or to stay, it needs to know if the player took a third card or not and, if so, the value of that card. If he did take a card, we denote its value $G(P + 5)$ by Z (see line 520), and if not, we let $Z = -1$ in line 540. The computer also needs to know the count of his down cards (hole cards). We use H for that purpose:

```
550   H=0
560   FOR J=P+2 TO P+4 STEP 2
570   H=H+G[J]-10*INT((H+G[J])/10)
580   NEXT J
```

We consult Table 3.1(b), and program the computer to follow the strategy that is laid out there:

```
590   IF H<3 OR (H=3 AND Z <> 8) THEN 690
600   IF H=4 AND (Z=-1 OR (Z>1 AND Z<8)) THEN 690
610   IF H=5 AND (Z=-1 OR (Z>3 AND Z<8)) THEN 690
620   IF H=6 AND Z=-1 AND RND(1)<851/2288 THEN 690
630   IF H=6 AND (Z=6 OR Z=7) THEN 690
640   PRINT "THE BANKER STAYS WITH"
650   FOR J=P+2 TO P+4 STEP 2
660   GOSUB 990
670   NEXT J
680   GOTO 380
690   PRINT "THE BANKER DRAWS THE"
700   J=P+N+3
710   GOSUB 990
720   PRINT "TO THE"
730   GOTO 650
```

Note how control passes back to line 380 for the next coup. There, P is advanced by N, and then N is set back to zero.

We are now ready to tackle the case where the operator is the banker and the computer is the active player. We could utilize the portions of the above program that pertain to the dealing of cards and use a variable, say T, with values 1 and 2, and the computed GOTO statement to pass control back to the pertinent part of the program. We won't do so, however, for it would not lead to an appreciable simplification of the program and might confuse the reader. Instead, we use

```
740   PRINT "THE BANKER IS DEALT"
750   FOR J=P+2 TO P+4 STEP 2
760   GOSUB 990
770   NEXT J
```

The active player (the computer) needs to know the count of his down cards to decide whether to stay or draw. We use G for that purpose:

```
780   G=0
790   FOR J=P+1 TO P+3 STEP 2
800   G=G+G[J]-10*INT((G+G[J])/10)
810   NEXT J
```

From Table 3.1(a), we have the following:

```
820   IF G>4 AND (G <> 5 OR RND(1)>9/11) THEN 870
830   PRINT "THE PLAYER DRAWS THE"
840   J=P+5
850   GOSUB 990
860   GOTO 880
870   PRINT "THE PLAYER STAYS."
880   PRINT "DO YOU WANT ANOTHER CARD ";
890   INPUT A$
900   IF A$[1,1]="N" THEN 940
910   PRINT "HERE IS THE"
```

```
920   J=P+N+3
930   GOSUB 990
940   PRINT "THE PLAYER IS SITTING ON"
950   FOR J=P+1 TO P+3 STEP 2
960   GOSUB 990
970   NEXT J
980   GOTO 380
```

We add

```
10   DIM A$[6],S[312],T[312],O[312],V[312],L[312],G[312]
20   DIM V$[65],S$[32]
```

```
1020   END
```

and we have a workable, though primitive, program. A sample run
follows. (Note: To terminate the run, you have to mash down the
BREAK-key, ATTN-key, or ESC-key. Otherwise, the run would go on
until all 312 cards are used up.)

```
STAND BY - THE CARDS ARE BEING SHUFFLED.
PLEASE CUT - BY ENTERING A NUMBER BETWEEN 1 AND 312.
?37
DO YOU WANT TO BE BANKER OR PLAYER ?PLAYER
KING    OF CLUBS
KING    OF CLUBS
DO YOU WANT ANOTHER CARD ?YES
HERE IS THE
TEN     OF DIAMONDS
THE BANKER DRAWS THE
NINE    OF HEARTS
TO THE
SEVEN OF CLUBS
FOUR    OF CLUBS
DO YOU WANT TO BE BANKER OR PLAYER ?PLAYER
QUEEN OF DIAMONDS
NINE    OF SPADES
DO YOU WANT ANOTHER CARD ?NO
THE BANKER DRAWS THE
FIVE    OF DIAMONDS
TO THE
NINE    OF CLUBS
FOUR    OF CLUBS
DO YOU WANT TO BE BANKER OR PLAYER ?BANKER
THE BANKER IS DEALT
FOUR    OF DIAMONDS
TWO     OF CLUBS
THE PLAYER STAYS.
DO YOU WANT ANOTHER CARD ?NO
THE PLAYER IS SITTING ON
TWO     OF DIAMONDS
FOUR    OF DIAMONDS
DO YOU WANT TO BE BANKER OR PLAYER ?BANKER
THE BANKER IS DEALT
SEVEN OF SPADES
SEVEN OF CLUBS
THE PLAYER STAYS.
DO YOU WANT ANOTHER CARD ?YES
```

```
HERE IS THE
KING  OF HEARTS
THE PLAYER IS SITTING ON
THREE OF SPADES
SIX   OF DIAMONDS
DO YOU WANT TO BE BANKER OR PLAYER ?BANKER
THE BANKER IS DEALT
EIGHT OF SPADES
ACE   OF SPADES
THE PLAYER STAYS.
DO YOU WANT ANOTHER CARD ?NO
THE PLAYER IS SITTING ON
SIX   OF CLUBS
JACK  OF HEARTS
DO YOU WANT TO BE BANKER OR PLAYER ?
STOP
```

The print-out of Fig. 3.1 was produced by a program that reflects, in essence, the rules and strategies for chemin-de-fer that were discussed in the preceding two sections. However, in order to keep the program from getting out of hand, we have introduced a number of restrictions.

There are only six players, one being the operator and the others being fictional characters. These fictional characters have been imbued to some degree with a will of their own: Their initial bid for the bank, their bets, and their choice of whether or not to go *banco* are selected, within reasonable limits, by means of the random function. Who is to play first is also determined by means of the random function.

The computer is programmed to play the optimal mixed strategy represented in Table 3.1(a) if it is the active player and the one in Table 3.1(b) if it is the banker.

If the operator is the banker, then the role of the active player is assigned (by the computer) to the nearest player, provided he still has money, to the left of the banker—unless somebody goes *banco*. In that case, the one who goes *banco* is the active player. If the computer is the banker, then the operator, of course, is the active player.

Only the operator's down cards (be he banker or player) are revealed after being dealt, but this knowledge, of course, is kept from the computer since it is playing the role of the opposition. The computer's down cards are not revealed until the end of the *coup*. You might say that the computer acts like a real schizo, knowing the operator's down cards as he does and at the same time not knowing them. It does indeed! There is, of course, nothing strange about this. A computer does not evaluate any information it may have stashed away in its bowels unless it is specifically programmed to do so. (At least, we hope that this is still the case!)

Our program does not allow the banker to put any winnings *en garage*. What is left of his winnings after the house has been paid its 5 percent and after the *chef de jeu* has received a tip—the amount to be chosen such that the remaining amount is rounded off to a multiple of 100—is to be added to the bank's holdings and is up for grabs.

No provisions are made for a *suivi*, and no provisions are made for a *banco avec la table* bet. (Both could be accommodated, however, with a modification of the program; see p. 107.)

The program comes to a halt not only when all but the operator are out of money but also when the operator is bankrupt, even though some of the other players may still be flush.

The core of the program—the shuffling, dealing, counting, and the computer's strategy, which has been explained in conjunction with the foregoing primitive program—is embodied, augmented by some frills, in the full program displayed in Fig. 3.2.

Specifically, you will find the shuffling of the cards in a subroutine, lines 2580 to 2930. Note how we "burned" the first three cards in line 2920 to add realism to the simulation. The reason for putting the shuffle into a subroutine is the same as in trente-et-quarante (see Chap. 1). We also want access to it when all but six cards (out of the 312) have been used up; see line 1130. The six decks are then reshuffled in response to the GOSUB 2550 command in line 1140.

The dealing of the cards is now to be found in lines 1160 to 1720 (if the computer is the banker) and in lines 2080 to 2490 (if the operator is the banker). The banker's strategy is to be found in lines 1430 to 1470, and the player's strategy, in line 2250.

The "frills" incorporated into our program not only concern the formatting of the output (by the liberal use of the TAB- and LIN-functions and, in one instance—see lines 590 and 600—by the use of a PRINT USING statement in conjunction with an IMAGE statement) but go well beyond such trivialities. A "natural" (*le petit* or *le grand*) is announced in lines 1270 and 1290 or 2210 and 2230, and the deal comes to a halt. The computer figures out who wins at the end of each *coup*. Unless there is a *coup neutre*, the winner is announced, and the computer keeps a running tab of the players' finances. [Player number J holds M(J) francs and bets B(J) francs. The banker wins or loses C francs, the sum of all bets.]

A number R (between 1 and 6) is chosen at random in line 310; the player listed in lines 210–230 in position R is assigned first place; and the operator is assigned the (7−R)th place. We leave it to the reader to unravel our method of picking the names of the players in their proper order from the string variable P$ (defined in lines 270 to 300), using the sub-string capabilities of our computer. We also leave it to the reader to figure out how the computer picks out the operator's surname in lines 160–180 from the input in line 130.

After everybody has been given a chance to bid for the bank, the computer—in the FOR-NEXT loop 500 to 550—figures out who was

```
10     PRINT TAB(23)"*****************"
20     PRINT TAB(23)"* CHEMIN-DE-FER *"
30     PRINT TAB(23)"*****************"
40     DIM N$[20],P$[132],M[12],B[6],A$[3],S[312],T[312]
50     DIM V[312],L[312],O[312],G[312],V$[65],S$[32],B$[17]
60     V$[1,30]="ACE    TWO    THREEFOUR FIVE SIX   "
70     V$[31,65]="SEVENEIGHTNINE TEN   JACK QUEENKING "
80     S$="CLUBS    DIAMONDSHEARTS  SPADES   "
90     GOSUB 2570
95     REM X ARE THE CASINO'S ACCUMULATED COMMISSIONS, Y THE
96     REM ACCUMULATED TIPS FOR THE PIT-BOSS
100    X=0
110    Y=0
120    PRINT "PLEASE IDENTIFY YOURSELF ";
130    INPUT N$
140    PRINT LIN(1)
150    L=LEN(N$)
155    REM LOOP 160 TO 180 LOOKS FOR THE BLANK SPACE BETWEEN
156    REM THE OPERATORS FIRST NAME AND SURNAME
160    FOR I=1 TO L
170    IF N$[I,I]=" " THEN 190
180    NEXT I
190    PRINT "BON SOIR, MONSIEUR "N$[I+1,L]", COMMENT ALLEZ-VOUS ";
200    PRINT "? THE OTHER PLAYERS"
210    PRINT "ARE SIR HILARY BRAY,LE MONSTRE DE LILLE,THE COMTE ";
220    PRINT "DE BLEUCHAMP,"
230    PRINT "COMMANDER BOND, AND SIR PALMER EPSOM-OAKS. YOU WILL";
240    PRINT ", IF YOU"
250    PRINT "PLEASE, PLAY IN THE FOLLOWING ORDER:"
260    PRINT LIN(1)
270    P$[1,44]="SIR HILARY LE MONSTRE M. LE COMTECMDR. BOND "
280    P$[45,58]="SIR PALMER M. "
290    P$[59,66]=N$[I+1,L]
300    P$[67,132]=P$
305    REM PLAYER NUMBER R IS PICKED OUT TO START THE BIDDING
306    REM FOR THE BANK
310    R=INT(RND(1)*6)+1
320    P$[1,67]=P$[11*R-10,11*R+56]
330    FOR J=1 TO 6
340    PRINT TAB(22);J;P$[1+(J-1)*11,J*11]
350    NEXT J
360    PRINT LIN(1)
370    FOR J=1 TO 6
380    PRINT P$[1+(J-1)*11,J*11]", HOW MUCH MONEY DO YOU WISH ";
390    PRINT "TO PUT UP ";
395    REM PLAYER NUMBER 7-R IS THE OPERATOR
400    IF J=7-R THEN 440
405    REM FICTIONAL PLAYERS PUT UP AMOUNTS M(J) BETWEEN
406    REM 1000 FRANCS AND 10000 FRANCS
410    M[J]=1000*INT(RND(1)*10)+1000
420    PRINT "?"M[J]
430    GOTO 480
440    INPUT M[J]
450    IF M[J]/1000=INT(M[J]/1000) THEN 480
460    PRINT "ONLY MULTIPLES OF THOUSAND FRANCS, PLEASE !"
470    GOTO 380
480    NEXT J
490    PRINT LIN(1)
495    REM THE LOOP 500 TO 550 PICKS THE PLAYER WHO WAS FIRST
496    REM TO PUT UP THE LARGEST AMOUNT
500    FOR K=1 TO 6
```

Fig. 3.2 Master program "Chemin-de-Fer"

```
510    FOR J=K+1 TO 6
520    IF M[K]<M[J] THEN 550
530    NEXT J
540    GOTO 560
550    NEXT K
560    PRINT P$[1+(K-1)*11,K*11]" HAS THE BANK FOR "M[K]" FRANCS."
570    PRINT LIN(1)
580    B$="*** BANCO DEMILLE"
590    PRINT   USING 600;B$[1,12],M[K]/1000,B$[13,17],B$[1,3]
600    IMAGE 17X,12A,X,2D.D,X,5A,X,3A
610    PRINT LIN(1)
620    PRINT TAB(12)"*** MESSIEURS, FAITES VOS JEUX... ***"
630    PRINT LIN(1)
640    C=0
650    FOR J=1 TO 6
660    IF J <> K AND M[J] <> 0 THEN 690
670    B[J]=0
680    GOTO 930
690    PRINT P$[1+11*(J-1),11*J]", ENTER YOUR BET !";
700    IF J=7-R THEN 760
705    REM THE FICTIONAL PLAYERS BET AMOUNTS B(J) BETWEEN 100
706    REM AND 1500 FRANCS, EXCEPT THAT OCCASIONALLY, ONE
707    REM GOES BANCO - SEE LINE 710
710    IF INT(RND(1)*100)=69 AND M[J] >= M[K] THEN 960
720    B[J]=((INT(RND(1)*15)*100+100) MIN (M[K]-C)) MIN M[J]
730    IF B[J]=M[K] THEN 960
740    PRINT B[J]
750    GOTO 920
760    INPUT B[J]
770    IF B[J]=M[K] AND B[J] <= M[J] THEN 960
780    IF B[J] <= M[J] OR B[J] <= M[K]-C THEN 820
790    PRINT "NEITHER YOU NOR THE BANK CAN COVER THAT BET."
800    PRINT "ENTER AN AMOUNT NOT TO EXCEED "(M[K]-C) MIN M[J];
810    GOTO 890
820    IF B[J] <= M[J] OR B[J]>M[K]-C THEN 860
830    PRINT "YOU DON'T HAVE ENOUGH MONEY. ENTER AN AMOUNT "
840    PRINT "NOT TO EXCEED "M[J];
850    GOTO 890
860    IF B[J]>M[J] OR B[J] <= M[K]-C THEN 920
870    PRINT "THE BANK CANNOT COVER THAT BET. ENTER AN AMOUNT"
880    PRINT "NOT TO EXCEED "M[K]-C;
890    IF M[J]<M[K] THEN 910
900    PRINT "OR GO BANCO ";
910    GOTO 760
920    C=C+B[J]
930    NEXT J
935    REM PLAYER NUMBER S IS THE ONE TO THE BANKER'S LEFT
940    S=K+5-6*INT((K+4)/6)
950    GOTO 1070
960    FOR I=1 TO 6
970    B[I]=0
980    NEXT I
990    B[J]=M[K]
1000   PRINT LIN(1)
1010   PRINT TAB(14)"*** BANCO - LE BANCO EST FAIT ***"
1020   PRINT LIN(1)
1030   PRINT P$[1+(J-1)*11,J*11]" BETS "M[K]" FRANCS ! ";
1040   PRINT "ALL OTHER BETS ARE VOID."
1050   C=M[K]
1060   S=J
1070   PRINT LIN(1)
1080   PRINT TAB(9)"** LES JEUX SONT FAITS, RIEN NE VA PLUS **"
```

Fig. 3.2 Master program "Chemin-de-Fer" (cont'd)

```
1090    PRINT LIN(1)
1095    REM N IS THE NUMBER OF CARDS USED IN ONE COUP, G IS
1096    REM THE COUNT OF THE PLAYER'S HAND, H THE COUNT OF
1097    REM THE BANKER'S HAND
1100    N=0
1110    G=0
1120    H=0
1130    IF P<307 THEN 1150
1135    REM IF ONLY SIX CARDS ARE LEFT, THE CARDS ARE RE-SHUFFLED
1140    GOSUB 2550
1150    IF K=7-R THEN 2080
1155    REM THE PLAYER IS DEALT TWO CARDS IN THE LOOP 1160 TO 1200
1160    FOR J=P+1 TO P+3 STEP 2
1170    PRINT TAB(38);
1180    GOSUB 2940
1190    G=G+G[J]-10*INT((G+G[J])/10)
1200    NEXT J
1210    PRINT
1215    REM THE BANKER'S HAND IS EVALUATED IN THE LOOP 1220 TO
1216    REM 1240
1220    FOR J=P+2 TO P+4 STEP 2
1230    H=H+G[J]-INT((H+G[J])/10)*10
1240    NEXT J
1245    REM THE DEALT HANDS ARE CHECKED FOR COUNTS OF 8 OR 9
1246    REM IN LINES 1250 TO 1300
1250    IF H<8 AND G<8 THEN 1310
1260    IF H=9 OR G=9 THEN 1290
1270    PRINT TAB(23)"*** LE PETIT ***"
1280    GOTO 1480
1290    PRINT TAB(23)"*** LE GRAND ***"
1300    GOTO 1480
1310    PRINT "DO YOU WANT ANOTHER CARD ";
1320    INPUT A$
1330    IF A$[1,1]="N" THEN 1420
1335    REM THE VALUE OF THE PLAYER'S THIRD CARD IS Z. IF HE STAYS,
1336    REM THEN Z IS SET TO -1
1340    Z=G[P+5]
1350    PRINT
1355    REM THE PLAYER IS DEALT A THIRD CARD IN LINES 1370 TO 1400
1360    PRINT "HERE IS THE "TAB(38);
1370    J=P+5
1380    GOSUB 2940
1390    PRINT
1400    G=G+G[P+5]-10*INT((G+G[P+5])/10)
1410    GOTO 1430
1420    Z=-1
1425    REM 1430 TO 1470 CONTAINS THE BANKER'S OPTIMAL STRATEGY
1430    IF H<3 OR (H=3 AND Z <> 8) THEN 1640
1440    IF H=4 AND (Z=-1 OR (Z>1 AND Z<8)) THEN 1640
1450    IF H=5 AND (Z=-1 OR (Z>3 AND Z<8)) THEN 1640
1460    IF H=6 AND Z=-1 AND RND(1)<851/2288 THEN 1640
1470    IF H=6 AND (Z=6 OR Z=7) THEN 1640
1480    PRINT LIN(1)
1490    PRINT "THE BANKER STAYS WITH "
1500    PRINT
1505    REM THE BANKER'S HAND IS REVEALED IN LINES 1510 TO 1530
1510    FOR J=P+2 TO P+4 STEP 2
1520    GOSUB 2940
1530    NEXT J
1540    PRINT LIN(1)
1550    IF H <> G THEN 1610
1560    PRINT LIN(1)
```

Fig. 3.2 Master program "Chemin-de-Fer" (cont'd)

```
1570    PRINT TAB(21)"*** COUP NEUTRE ***"
1580    PRINT LIN(1)
1590    P=P+N
1600    GOTO 1100
1610    PRINT "FOR A COUNT OF"H"TO THE OPPOSITION'S COUNT OF";
1620    PRINT G"TO"
1630    GOTO 1720
1640    PRINT "THE BANKER DRAWS THE "
1650    PRINT
1660    J=N+P+3
1670    H=H+G[P+N+3]-10*INT((H+G[P+N+3])/10)
1680    GOSUB 2940
1690    PRINT
1700    PRINT "TO THE"
1710    GOTO 1500
1720    P=P+N
1725    REM IN 1730, THE COMPUTER CHECKS ON WHO WON
1730    IF H<G THEN 1890
1735    REM M IS THE TIP FOR THE PIT-BOSS
1740    M=M[K]+95*C/100-100*INT((M[K]+95*C/100)/100)
1750    PRINT "WIN"C"FRANCS."5*C/100"FRANCS GO INTO THE ";
1760    PRINT "CAGNOTTE"
1770    IF M=0 THEN 1800
1780    PRINT "AND"M"FRANCS ARE 'UN POURBOIRE' FOR THE CHEF DE ";
1790    PRINT "JEU."
1800    X=X+5*C/100
1810    Y=Y+M
1820    PRINT
1825    REM THE BANKER GETS 95*C/100-M FRANCS OUT OF HIS WINNINGS
1826    REM OF C FRANCS
1830    M[K]=M[K]+95*C/100-M
1840    FOR J=1 TO 6
1850    IF J=K THEN 1870
1855    REM EACH PLAYER LOSES HIS BET OF B(J) FRANCS IN LINE 1860
1860    M[J]=M[J]-B[J]
1870    NEXT J
1880    GOTO 1950
1890    PRINT "LOSE"C"FRANCS.THE BANK PASSES TO THE NEXT PLAYER."
1895    REM THE BANKER LOSES C FRANCS
1900    M[K]=M[K]-C
1910    FOR J=1 TO 6
1920    IF J=K THEN 1940
1925    REM EACH PLAYER WINS HIS BET OF B(J) FRANCS IN LINE 1930
1930    M[J]=M[J]+B[J]
1940    NEXT J
1950    PRINT
1960    PRINT "THE PLAYERS ARE NOW HOLDING THE FOLLOWING AMOUNTS:"
1970    PRINT
1980    FOR J=1 TO 6
1990    PRINT TAB(20);P$[1+(J-1)*11,J*11]" ... "M[J]
2000    NEXT J
2010    PRINT LIN(1)
2020    IF H>G THEN 2040
2025    REM IN LINES 2030,2040, THE CLOSEST PLAYER TO THE BANKER'S
2026    REM RIGHT, WHO STILL HAS MONEY, IS DETERMINED SO THAT HE
2027    REM MAY TAKE THE BANK
2030    K=K+1-6*INT(K/6)
2040    IF M[K]=0 THEN 2030
2045    REM IF THE CONDITION IN LINE 2050 IS SATISFIED, THE BANKER
2046    REM IS THE ONLY ONE WHO HAS SOME MONEY LEFT AND THE GAME IS
2047    REM TERMINATED
2050    IF M[1]+M[2]+M[3]+M[4]+M[5]+M[6]=M[K] THEN 3000
```

*Fig. 3.2 Master program "Chemin-de-Fer" (cont'd)*

```
2055   REM IF THE OPERATOR (PLAYER NUMBER 7-R) IS BANKRUPT, THE
2056   REM GAME TERMINATES
2060   IF M[7-R]=0 THEN 2980
2070   GOTO 560
2075   REM THE OPERATOR IS BANKER. HIS HAND IS DEALT AND EVALUATED
2076   REM IN LINES 2080 TO 2110
2080   FOR J=P+2 TO P+4 STEP 2
2090   GOSUB 2940
2100   H=H+G[J]-10*INT((G[J]+H)/10)
2110   NEXT J
2120   PRINT LIN(1)
2125   REM THE PLAYER'S HAND IS EVALUATED IN LINES 2130 TO 2150
2130   FOR J=P+1 TO P+3 STEP 2
2140   G=G+G[J]-10*INT((G[J]+G)/10)
2150   NEXT J
2155   REM IN 2160, 2170, THE CLOSEST PLAYER TO THE BANKER'S
2156   REM LEFT, WHO STILL HAS MONEY, IS DETERMINED TO BECOME
2157   REM ACTIVE PLAYER (PLAYER NUMBER S)
2160   IF M[S]>0 THEN 2190
2170   S=S+5-6*INT((S+4)/6)
2180   GOTO 2160
2185   REM THE DEALT HANDS ARE CHECKED FOR COUNTS OF 8 OR 9
2190   IF H<8 AND G<8 THEN 2250
2200   IF H=9 OR G=9 THEN 2230
2210   PRINT TAB(23)"*** LE PETIT ***"
2220   GOTO 2440
2230   PRINT TAB(23)"*** LE GRAND ***"
2240   GOTO 2440
2245   REM LINE 2250 CONTAINS THE PLAYER'S OPTIMAL STRATEGY
2250   IF G>4 AND (G <> 5 OR RND(1)>9/11) THEN 2330
2255   REM THE PLAYER DRAWS AN EXTRA CARD IN LINES 2260 TO 2300
2260   PRINT P$[1+11*(S-1),11*S]" DRAWS THE ";
2270   PRINT TAB(38);
2280   J=P+5
2290   GOSUB 2940
2300   G=G+G[P+5]-10*INT((G+G[P+5])/10)
2310   GOTO 2340
2320   PRINT
2330   PRINT P$[1+(S-1)*11,S*11]" STAYS."
2340   PRINT
2350   PRINT P$[1+(K-1)*11,K*11]" DO YOU WANT ANOTHER CARD ";
2360   INPUT A$
2370   PRINT
2380   IF A$[1,1]="N" THEN 2450
2385   REM THE BANKER (OPERATOR) GETS AN EXTRA CARD IN
2386   REM LINES 2390 TO 2430
2390   PRINT "HERE IS THE "
2400   PRINT
2410   J=P+N+3
2420   H=H+G[P+N+3]-10*INT((H+G[P+N+3])/10)
2430   GOSUB 2940
2440   PRINT
2445   REM THE PLAYER'S HAND IS REVEALED IN LINES 2450
2446   REM TO 2490
2450   PRINT P$[1+11*(S-1),11*S]" IS SITTING ON ";
2460   FOR J=P+1 TO P+3 STEP 2
2470   PRINT TAB(38);
2480   GOSUB 2940
2490   NEXT J
2500   PRINT LIN(1)
2510   IF H=G THEN 1560
2520   PRINT "FOR A TOTAL COUNT OF"G", WHILE YOUR COUNT ";
```

*Fig. 3.2 Master program "Chemin-de-Fer" (cont'd)*

```
2530    PRINT "IS"H"AND YOU"
2540    GOTO 1720
2550    PRINT LIN(1)
2560    PRINT TAB(19)"*** LES CARTES PASSENT ***"
2570    PRINT LIN(1)
2575    REM THE SIX DECKS OF CARDS ARE SHUFFLED, FIRST EACH
2576    REM DECK INDIVIDUALLY, AND THEN ALL DECKS TOGETHER
2580    PRINT TAB(8)"*** STAND BY - ";
2590    PRINT "THE CARDS ARE BEING SHUFFLED ***"
2600    PRINT LIN(1)
2610    FOR W=0 TO 5
2620    FOR J=1+52*W TO 52+52*W
2630    S[J]=J
2640    NEXT J
2650    FOR I=52+52*W TO 2+52*W STEP -1
2660    J=INT(RND(1)*I)+1
2670    T=S[J]
2680    S[J]=S[I]
2690    S[I]=T
2700    NEXT I
2710    NEXT W
2720    FOR I=312 TO 2 STEP -1
2730    J=INT(RND(1)*I)+1
2740    T=S[J]
2750    S[J]=S[I]
2760    S[I]=T
2770    NEXT I
2780    PRINT TAB(2)"*** PLEASE CUT - BY ENTERING A NUMBER ";
2790    PRINT "BETWEEN 1 AND 312 ***"
2800    PRINT LIN(1)
2805    REM THE CARDS ARE CUT AND THEIR FACE VALUE V(J), SUIT
2806    REM S(J), AND GAME VALUE G(J) ARE DETERMINED
2810    INPUT Z
2820    FOR J=1 TO 312
2830    T[J]=S[J]+Z-312*INT((S[J]+Z-1)/312)
2840    NEXT J
2850    FOR J=1 TO 312
2860    S[J]=T[J]-52*INT((T[J]-1)/52)
2870    V[J]=S[J]-13*INT((S[J]-1)/13)
2880    L[J]=INT((S[J]-1)/13)
2890    O[J]=V[J] MIN 10
2900    G[J]=O[J]-10*INT(O[J]/10)
2910    NEXT J
2915    REM THE FIRST THREE CARDS ARE BURNED IN LINE 2920
2920    P=3
2930    RETURN
2935    REM A CARD IS DEALT IN LINES 2940,2950 AND COUNTED
2936    REM IN LINE 2960
2940    PRINT "*** "V$[1+5*(V[J]-1),5+5*(V[J]-1)]" OF ";
2950    PRINT S$[1+8*L[J],8+8*L[J]]" ***"
2960    N=N+1
2970    RETURN
2980    PRINT "YOU WERE TAKEN TO THE CLEANERS BY EXPERTS."
2990    GOTO 3070
3000    IF K=7-R THEN 3040
3010    PRINT "SINCE NOBODY BUT "P$[1+(K-1)*11,K*11]" HAS ANY ";
3020    PRINT "MONEY LEFT,"
3030    GOTO 3060
3040    PRINT "YOU CLEANED THEM OUT GOOD!NOBODY BUT YOU ";
3050    PRINT "HAS ANY MONEY LEFT AND"
3060    PRINT "THE GAME COMES TO A SCREECHING HALT !"
```

*Fig. 3.2 Master program "Chemin-de-Fer" (cont'd)*

```
3070  PRINT LIN(1)
3080  PRINT 'THE CASINO MADE'X'FRANCS IN COMMISSIONS AND THE'
3090  PRINT 'CHEF DE JEU MADE'Y'FRANCS IN TIPS.'
3100  END
```

*Fig. 3.2 Master program "Chemin-de-Fer" (cont'd)*

first to bid the highest amount and makes him banker. In our program, the banker's place in the order of players is denoted by K (see line 560).

When the operator is the banker, somebody has to be designated as the active player. We make player number S the active player, S being chosen in line 940 (the player to the banker's left), unless somebody has gone *banco*. In that case, the one who has gone *banco* is made active player in line 1060. However, if nobody has gone *banco* and the player to the banker's left is bankrupt, the role is passed on in the clockwise direction (see lines 2160 to 2180).

Finally, let us note that $M(J) \geq 0$ for all $J$ = 1, 2, 3, 4, 5, or 6, and therefore, $M(1)+M(2)+M(3)+M(4)+M(5)+M(6) = M(K)$ if and only if $M(J) = 0$ for all $J \neq K$, that is, nobody but the banker has any money left. We use this criterion in line 2050 to bring the game to a halt. (Note that if the operator is out of money, the game is brought to a halt in line 2060.)

## MODIFICATIONS OF THE COMPUTER PROGRAM

Our program is designed to accommodate six players: the operator and five fictional characters. It is a simple matter to extend it to nine (or any number of) players by augmenting the string variable P$ in lines 270 and 280; changing the pertinent FOR-NEXT loops from FOR J = 1 TO 6 to FOR J = 1 TO 9 (or whatever) in lines 330, 370, 500, 650, and 1980; changing line 510 from FOR J = K+1 TO 6 to FOR J = K+1 TO 9; changing line 310 to R = INT(RND(1)*9)+1; and replacing $7-R$ by $10-R$ wherever it occurs. Also note that the program, as it stands now, only accommodates players whose names, including the appellation, do not exceed 11 characters. The reader can easily change this by making the appropriate adjustments in lines 210 and 300 and by replacing the factor 11 in P$(1+(J−1)*11,J*11) by the new number of characters wherever this particular sub-string occurs.

While the program does not contain any explicit instructions as to the admissible magnitude of the initial bid for the bank, the fictional players are programmed to bid (at random) amounts anywhere from 1,000 francs to 10,000 francs (see line 410). If the operator is a reasonable man and also stays within these limits, then the bank cannot ever exceed 60,000 francs (90,000 francs for nine players), and the IMAGE statement in line 600 is adequate. Should larger amounts be permitted, then the 2D.D (which formats the output to two digits before and one digit after the decimal point) will have to be changed to 3D.D, or even 4D.D, depending on how high one wishes to go. Since we are not playing for real money anyway, we suggest that there is not much point to such changes.

No restrictions, other than the operator's holdings and the amount left in the bank, are placed on the operator's bets. The fictional players, however, are programmed to bet amounts between 100 francs and 1500 francs (the specific amount being chosen at random), but not to exceed their holdings nor the amount that is left in the bank (see line 720).

Every once in a hundred whiles (or even less frequently), a fictional character will go *banco* (see line 710). You may increase or decrease the relative frequency of *banco* bets by changing the parameters in line 710. For example, IF INT(RND(1)*10) = 5 AND M(J) >= M(K) THEN 960 will increase the frequency about tenfold. The operator in his capacity as player may, of course, go *banco* any time he wishes, provided he has enough money.

If the computer is the banker, then there is not much point in designating anybody but the operator as the active player. However, if the operator is the banker, then we could designate the highest bidder as the active player rather than the nearest player to the operator's left who still has money left, as we did in our program. We may do this with a device that is analogous to the one we used in figuring out who put in the first highest bid for the bank in lines 500 to 550. Here is what we will do (for six players):

```
940
1060
1091   FOR W=1 TO 6
1092   FOR J=W+1 TO 6
1093   IF B[W]<B[J] THEN 1096
1094   NEXT J
1095   GOTO 1097
1096   NEXT W
1097   S=W
2160
2170
2180
```

(Note that lines 940, 1060, 2160, 2170, and 2180 are now irrelevant. Since player number W entered the highest bet, he has to have some money left!)

We may change our banking rules as follows: The banker puts his winnings (minus the house take and tip) *en garage*, that is, he withdraws it from the bank and, in effect, from the game. We may keep track of each player's "garage" by means of a vector, Q(J):

```
31   DIM Q[6]
32   MAT Q=ZER[6]

1830   Q[K]=Q[K]+95*C/100-M

2001   PRINT
2002   PRINT "AND HAVE THE FOLLOWING AMOUNTS 'EN GARAGE':"
2003   PRINT
2004   FOR J=1 TO 6
2005   PRINT TAB(20);P$[1+(J-1)*11,J*11]" ... "Q[J]
2006   NEXT J
```

Here is an excerpt from a computer run based on our program in Fig. 3.2 with the above modifications:

```
THE PLAYERS ARE NOW HOLDING THE FOLLOWING AMOUNTS:

                    M. LE COMTE  ...   5600
                    CMDR. BOND   ...   6000
                    SIR PALMER   ...   9300
                    M. SAGAN     ...   7500
                    SIR HILARY   ...   1200
                    LE MONSTRE   ...   7100

AND HAVE THE FOLLOWING AMOUNTS 'EN GARAGE':

                    M. LE COMTE  ...   0
                    CMDR. BOND   ...   0
                    SIR PALMER   ...   0
                    M. SAGAN     ...   7000
                    SIR HILARY   ...   0
                    LE MONSTRE   ...   4700


LE MONSTRE  HAS THE BANK FOR  7100     FRANCS.
```

We may let it go at that, or we may let the player who runs out of active money withdraw the funds from his "garage." We leave the details to the reader.

We may wish to change the strategy to simulate a game where the banker has to stay on 6 if the player stays. To optimize the player's strategy in that case, we have to make him stay or draw on 5, with probability ½ each. Here is how we do it:

```
1460
2250   IF G > 4 AND (G<>5 OR RND(1) > .5) THEN 2330
```

(Note that the elimination of 1460 does the trick of changing the banker's strategy to the new one: It automatically makes him stay on 6 if $Z = -1$, that is, if the player stays.)

Next, let's take care of a *banco suivi* bet after a player lost a *banco* bet and has enough money left to go *banco suivi*. First of all, we have to keep track of who went *banco* and lost:

```
115  B = O
1041  B = J
1901  B = O
```

(The banker lost in lines 1890 and 1900, and the question about a *banco suivi* bet does not arise. Hence, B is set back to 0. It is all right to use B as a variable even though it is also used to designate the "betting vector" B(J). The computer knows how to keep them apart.)

If the unsuccessful *banco* bettor still has at least as much money as there is in the bank, we let the computer ask him if he wants to go *banco suivi* and proceed accordingly:

```
630   IF B=0 THEN 643
631   IF M[B]<M[K] THEN 643
632   PRINT P$[1+(B-1)*11,B*11]",DO YOU WANT TO GO BANCO SUIVI ";
633   IF B=7-R THEN 637
634   IF RND(1)<2/3 THEN 642
635   PRINT "YES"
636   GOTO 640
637   INPUT A$
638   IF A$[1,1]="Y" THEN 640
639   GOTO 643
640   J=B
641   GOTO 960
642   PRINT "NO"
643   B=0
644   C=0
```

To introduce a note of surprise into our game, we let a player (other than the operator) who *went* banco and lost go *banco suivi* with probability $\frac{1}{3}$—if he has enough money, that is (see line 634). We leave it to the reader to accommodate a *banco avec la table* bet. Such a bet, B(B), has to be equal to what is left of M(K) after all the other bets, B(J) for J $\neq$ B, have been subtracted. We suggest letting number B place his bet *after* all other bets have been placed.

Finally, let us mention that we can keep the game going as long as two players have any money left, even if the operator is out of it. This change is accomplished simply by eliminating 2060, 2980, and 2990. (Note: Unless further modifications are made, the operator is still called upon to perform as an active player.)

# Chapter 4

# craps

```
********
* CRAPS *
********
```

WHEN ASKED TO 'NAME YOUR BET', ENTER ONE OF THE FOLLOWING:

```
        PASS        (PAY-OFF 1:1)
        DON'T PASS  (PAY-OFF 1:1)
        FIELD       (PAY-OFF 1:1, 2:1 ON 2,12)
        BIG SIX     (PAY-OFF 1:1)
        BIG EIGHT   (PAY-OFF 1:1)
        ANY CRAPS   (PAY-OFF 7:1)
        7           (PAY-OFF 4:1)
        11          (PAY-OFF 14:1)
        3           (PAY-OFF 14:1)
        2           (PAY-OFF 29:1)
        12          (PAY-OFF 29:1)
        HARD WAY    (PAY-OFF 9:1 ON 6,8 AND 7:1 ON 4,10)
```

```
    *** SHOOTER MUST WAGER ON PASS OR DON'T PASS ***
```

PLEASE IDENTIFY YOURSELF ?HANS SAGAN

HELLO SLICK HANS . MEET THE OTHER CHARACTERS WHO
CROWD AROUND THE CRAP TABLE: FRIENDLY-FRIENDLY, PADDY
THE CORK, CIEL MAGFIL, CHICAGO SAM, AND LITTLE JIM.

HOW MUCH MONEY DO YOU HAVE ?1000
YOU WON'T HAVE IT FOR LONG !

```
        *** FIRST HIGH ROLL GETS TO SHOOT FIRST ***

FRIENDLY-FRIENDLY ROLLS   *
                      *          *
                      *
```

*Fig. 4.1 Run of the program "Craps"*

109

```
PADDY THE CORK     ROLLS                 *
                             *
                                            *

CIEL MAGFIL        ROLLS   * *         *
                                         *
                           * *           *

CHICAGO SAM        ROLLS   * *       * *
                            *
                           * *       * *

LITTLE JIM         ROLLS   * *       *
                            *           *
                           * *           *

SLICK HANS         ROLLS   * *       *

                           * *         *
```

          *** CHICAGO SAM        IS COMING OUT ***

SLICK HANS , NAME YOUR BET ?PASS

HOW MUCH DO YOU WANT TO BET ?10

ON THE COME-OUT ROLL, CHICAGO SAM         ROLLS

```
                    ***         * *
                                 *
                    ***         * *
```

YOU WON $  10   AND YOU NOW HAVE $  1010

               *** CHICAGO SAM        IS COMING OUT ***

SLICK HANS , NAME YOUR BET ?DON'T PASS

HOW MUCH DO YOU WANT TO BET ?60

ON THE COME-OUT ROLL, CHICAGO SAM         ROLLS

```
                 *         *
                  *         *
                   *         *
```

POINT  6

DO YOU WANT TO LAY SINGLE ODDS AT A PAY-OFF OF   5      TO 6?YES

NOW COME THE POINT ROLLS *        ***

```
                       *        ***

                                * *
                       *         *
                                * *
```

Fig. 4.1 Run of the program "Craps" (cont'd)

POINT !

YOU LOST $ 120   AND YOU NOW HAVE $  890


              *** CHICAGO SAM        IS COMING- OUT ***

SLICK HANS , NAME YOUR BET ?PASS

HOW MUCH DO YOU WANT TO BET ?90


ON THE COME-OUT ROLL, CHICAGO SAM        ROLLS
```
                    *         * *
                             *
                    *         * *
```
YOU WON $  90    AND YOU NOW HAVE $  980


              *** CHICAGO SAM        IS COMING OUT ***

SLICK HANS , NAME YOUR BET ?FIELD

HOW MUCH DO YOU WANT TO BET ?90


ON THE COME-OUT ROLL, CHICAGO SAM        ROLLS
```
                 * *        *
                  *          *
                 * *          *
```
YOU LOST $  90   AND YOU NOW HAVE $  890

POINT   8

NOW COME THE POINT ROLLS *          *
```
                   *          *

                 * *        * *

                 * *        * *
```
POINT !


              *** CHICAGO SAM        IS COMING OUT ***

SLICK HANS , NAME YOUR BET ?BIG SIX

HOW MUCH DO YOU WANT TO BET ?90


ON THE COME-OUT ROLL, CHICAGO SAM        ROLLS

*Fig. 4.1 Run of the program "Craps" (cont'd)*

```
                                    * *
                          *
                                    * *

KEEP TRYING               *         * *
                                      *
                          *         * *

CHICAGO SAM       SEVENS OUT AND
LITTLE JIM        IS THE NEXT SHOOTER.

YOU LOST $  90   AND YOU NOW HAVE $   800


          *** LITTLE JIM         IS COMING OUT ***

SLICK HANS , NAME YOUR BET ?BIG EIGHT

HOW MUCH DO YOU WANT TO BET ?50

ON THE COME-OUT ROLL, LITTLE JIM          ROLLS

                          *         ***

                          *         ***

YOU WON $  50    AND YOU NOW HAVE $   850


          *** LITTLE JIM         IS COMING OUT ***

SLICK HANS , NAME YOUR BET ?ANY CRAPS

HOW MUCH DO YOU WANT TO BET ?100

ON THE COME-OUT ROLL, LITTLE JIM          ROLLS

                          *
                            *         *
                            *

YOU LOST $  100  AND YOU NOW HAVE $   750


POINT   4

NOW COME THE POINT ROLLS * *         * *

                          * *         * *

                          * *         *

                          * *          *

                          *           *

                            *           *
```

Fig. 4.1 Run of the program "Craps" (cont'd)

```
POINT !
              *** LITTLE JIM         IS COMING OUT ***


SLICK HANS , NAME YOUR BET ?7
HOW MUCH DO YOU WANT TO BET ?10
ON THE COME-OUT ROLL, LITTLE JIM          ROLLS
                        *         * *
                          *
                          *       * *
YOU WON $   40    AND YOU NOW HAVE $   790


              *** LITTLE JIM         IS COMING OUT ***


SLICK HANS , NAME YOUR BET ?11
HOW MUCH DO YOU WANT TO BET ?90

ON THE COME-OUT ROLL, LITTLE JIM          ROLLS

                            *
                        *
                            *
YOU LOST $   90    AND YOU NOW HAVE $   700


              *** LITTLE JIM         IS COMING OUT ***


SLICK HANS , NAME YOUR BET ?3
HOW MUCH DO YOU WANT TO BET ?100

ON THE COME-OUT ROLL, LITTLE JIM          ROLLS
                        * *       ***

                        * *       ***
YOU LOST $   100   AND YOU NOW HAVE $   600


POINT   10
NOW COME THE POINT ROLLS *         *
                          *
                            *         *
                          *       * *
                          *
                            *       * *
```

Fig. 4.1 Run of the program "Craps" (cont'd)

```
LITTLE JIM        SEVENS OUT AND
SLICK HANS        IS THE NEXT SHOOTER.


        *** SLICK HANS          IS COMING OUT ***


DO YOU WANT TO BET ON 'PASS' OR 'DON'T PASS' ?DON'T PASS

HOW MUCH DO YOU WANT TO BET ?30


ON THE COME-OUT ROLL, SLICK HANS          ROLLS

                    * *      *
                     *        *
                    * *        *

POINT   8

DO YOU WANT TO LAY SINGLE ODDS AT A PAY-OFF OF   5       TO 6?YES

NOW COME THE POINT ROLLS * *
                     *          *
                    * *
                    ***      * *

                    ***      * *

                              * *
                     *
                              * *

                   *          * *

                     *        * *
                              ***
                   *          ***

SLICK HANS         SEVENS OUT AND
FRIENDLY-FRIENDLY IS THE NEXT SHOOTER.

YOU WON $  55   AND YOU NOW HAVE $   655


        *** FRIENDLY-FRIENDLY IS COMING OUT ***


SLICK HANS , NAME YOUR BET ?2

HOW MUCH DO YOU WANT TO BET ?55


ON THE COME-OUT ROLL, FRIENDLY-FRIENDLY ROLLS

                    ***      ***

                    ***      ***
```

Fig. 4.1 Run of the program "Craps" (cont'd)

```
YOU LOST $  55   AND YOU NOW HAVE $  600

          *** FRIENDLY-FRIENDLY IS COMING OUT ***

SLICK HANS , NAME YOUR BET ?3

HOW MUCH DO YOU WANT TO BET ?100

ON THE COME-OUT ROLL, FRIENDLY-FRIENDLY ROLLS

                    *         ***

                      *       ***

YOU LOST $  100   AND YOU NOW HAVE $  500

POINT  8

NOW COME THE POINT ROLLS * *
                          *        *
                         * *

                         * *      ***
                          *
                         * *      ***

                         ***       *

                         ***        *

POINT !

          *** FRIENDLY-FRIENDLY IS COMING OUT ***

SLICK HANS , NAME YOUR BET ?12

HOW MUCH DO YOU WANT TO BET ?100

ON THE COME-OUT ROLL, FRIENDLY-FRIENDLY ROLLS

                    * *       * *

                    * *       * *

YOU LOST $  100   AND YOU NOW HAVE $  400

POINT  8

NOW COME THE POINT ROLLS *          *

                          *          *

                         *          *
                                      *
                          *          *
```

Fig. 4.1 Run of the program "Craps" (cont'd)

```
FRIENDLY-FRIENDLY SEVENS OUT AND
PADDY THE CORK    IS THE NEXT SHOOTER.


        *** PADDY THE CORK     IS COMING OUT ***


SLICK HANS , NAME YOUR BET ?HARD WAY

HOW MUCH DO YOU WANT TO BET ?100
ON 6, 8, 4, OR 10 ?6


ON THE COME-OUT ROLL, PADDY THE CORK     ROLLS
                ***
                            *
                ***
YOU LOST $  100   AND YOU NOW HAVE $   300


        *** PADDY THE CORK     IS COMING OUT ***


SLICK HANS , NAME YOUR BET ?HARD WAY

HOW MUCH DO YOU WANT TO BET ?10
ON 6, 8, 4, OR 10 ?8


ON THE COME-OUT ROLL, PADDY THE CORK     ROLLS
                            *
                    *       *
                            *

LET'S TRY AGAIN       * *     * *
                              *
                      * *     * *

                      * *     * *

                      * *     * *
```

Fig. 4.1 Run of the program "Craps" (cont'd)

```
YOU WON $  90   AND YOU NOW HAVE $  390


              *** PADDY THE CORK    IS COMING OUT ***


SLICK HANS , NAME YOUR BET ?HARD WAY

HOW MUCH DO YOU WANT TO BET ?190
ON 6, 8, 4, OR 10 ?4


ON THE COME-OUT ROLL, PADDY THE CORK    ROLLS

                                  *
                      *
                                    *

LET'S TRY AGAIN                   ***
                      *
                                  ***


PADDY THE CORK    SEVENS OUT AND
CIEL MAGFIL       IS THE NEXT SHOOTER.


YOU LOST $  190  AND YOU NOW HAVE $   200


              *** CIEL MAGFIL       IS COMING OUT ***


SLICK HANS , NAME YOUR BET ?HARD WAY

HOW MUCH DO YOU WANT TO BET ?200
ON 6, 8, 4, OR 10 ?10


ON THE COME-OUT ROLL, CIEL MAGFIL      ROLLS

                                  * *
                      *
                                  * *

LET'S TRY AGAIN                   ***
                                    *
                      ***


CIEL MAGFIL       SEVENS OUT AND
CHICAGO SAM       IS THE NEXT SHOOTER.


YOU LOST $  200  AND YOU NOW HAVE $   0


YOU GOT CLEANED OUT REAL GOOD !

DONE
```

*Fig. 4.1 Run of the program "Craps" (cont'd)*

*"You don't have to be a College Professor
to play a good game."*[1]
TOM JAMES

## HOW THE GAME IS PLAYED

Craps, or more accurately bank-craps, is played on a crap table
(see Fig. 4.2), the markings of which seem like a numerological
nightmare to the novice but make sense to the initiated—of which
there are many. ("Friendly" craps may be—and, as a matter of fact,
is—played almost anywhere: from the cement floor of the boiler room
in a New York tenement house to the floor of a tastelessly furnished,
newly rich electrician's recreation room. More will be said about
"friendly" craps shortly.) The chance event one bets on, a number
between 2 and 12, is generated by a pair of dice.

The players place their stakes on the field representing the chance
they wish to bet on, and the *shooter* (the player who has the dice) rolls
the dice with such vehemence that they bounce off the backboard at
the opposite end of the table.

The game is presided over by the *boxman*, who is in charge of
the money box and, with suspicious eyes, watches everything and
everybody. (Nowadays, the boxman quite often turns out to be a
boxwoman. Should we say "boxperson"?) The *stickman* handles the
dice, shoving them to and fro with a long stick. He also acts as barker
and announces the total after each roll. Two *dealers* collect losing
wagers and pay out winnings. They also give friendly advice (at least,
they are supposed to).

Somebody has to start rolling ("shooting") the dice. It may be the
first player to appear at the table or a player selected from a group of
players by intimidation or other means (for example, high roll shoots
first). Prior to each roll, players (in bank-craps, the shooter is also a
player and, like the rest of them, bets against the house) place their
bets. What to bet on and how to place the bets will be explained in the
next section. The shooter takes the dice, rattles them in his cupped
hands, maybe spits on them, and, finally, accompanied by ritualistic
incantations (such as "Baby needs new shoes") rolls them. Yelling is
discouraged but does occur on occasion.

The first roll is the *come-out roll*. It initiates a series of rolls that
may be terminated in three different ways:

1. If the come-out roll produces a 7 or 11, the series is terminated,

---

[1] Ref. [7], p. 87.

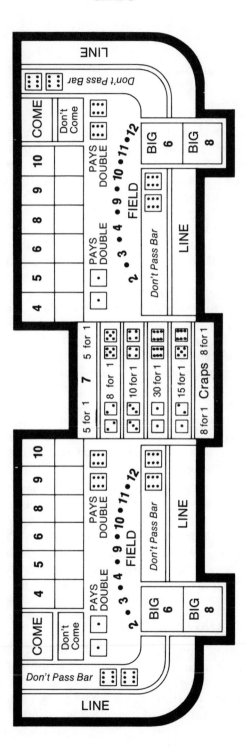

Fig. 4.2 Crap table

and the next roll is again a come-out roll. (The come-out roll "wins.")

2. If the come-out roll produces a 2, 3, or 12 *(craps)*, the series is terminated, and the next roll is again a come-out roll. (The come-out roll "loses.")

(Note that in both these cases, the series consists of one roll only. Since the probability of rolling 7 is 6/36, the probability of rolling 11 is 2/36, and the probabilities of rolling 2, 3, or 12 are 1/36, 2/36, and 1/36, respectively, it appears that the probability for a series to terminate with one roll is 12/36, or 1/3—that is, about one-third of all come-out rolls are terminal (see Appendix A, section 5).

3. If the come-out roll produces a number other than those mentioned—namely a 4, 5, 6, 8, 9, or 10—it is called a *point*, and the next roll is called a *point roll*. If the point roll produces a 7, the shooter *sevens out* and must surrender the dice to the next player in line, thus terminating the series. The next roll is again a come-out roll. If the point roll produces the point that has been established in the come-out-roll, the series also terminates and the next roll is a come-out-roll, but the shooter retains the dice. If neither of these events occur, the next roll is again a point-roll on which the shooter may seven out, roll the point that was established on the come-out roll, or neither, and so on. Mathematically speaking, this series is an *absorbing Markov chain* that terminates after finitely many rolls (see also Appendix A, section 6). As a matter of fact, it can be shown that the average number of rolls in a series—from come-out roll to final disposition—is 739/220, or 3.3590.

Observe that the 7, which is the number with the greatest probability of appearing, namely 1/6, plays a dual role. It is a "good" (winning) number on the come-out roll, but it is a "bad" (losing) number on a point roll. The terms "winning" and "losing," however, when used in this context, are to be taken with a grain of salt. A winning number may be a losing number to those who bet against it, and vice versa. More about this in the next section.

## WHAT TO BET AND HOW TO GO ABOUT IT

In bank-craps, you may place two kinds of bets: (1) those that have to be placed prior to the come-out-roll and which are decided at the termination of a series of point rolls if the come-out roll neither wins nor loses, and (2) those that may be placed prior to any roll.

The bets of the first kind are called *pass* and *don't pass*.

### Pass

> *Pay-off:*      Even money
> *Odds for winning:*      244:251
> *Expected loss:*      $14.14 per $1000 bet

You place your stake on the *pass line* as indicated by (A) in Fig. 4.3.

You win (1) if the "come-out roll wins," that is, if it produces a 7 or an 11 (which also terminates this particular series of rolls), or (2) if the "point roll wins," that is, if the come-out roll establishes a point (4, 5, 6, 8, 9, or 10) and this point is made on a subsequent point roll prior to an appearance of the 7. You lose (1) if the "come-out roll loses," that is, if it produces a 2, 3, or 12 (the shooter *craps out*), or, (2) if the "point roll loses," that is, if 7 appears on a roll before an established point is made.

As soon as a point is established, the dealer places a round *point marker* in the appropriate *point box*. (The point boxes are the fields that are marked 4, 5, 6, 8, 9, and 10 to either side of the center of the lay-out in Figs. 4.2 and 4.3.) When a shooter sevens out or when a point is made, the point marker is removed.

### Don't Pass (Bar Twelve)

> *Pay-off:*      Even money
> *Odds for winning:*      217:223
> *Expected loss:*      $13.63 per $1000 bet

You place your stake on the *don't pass (bar twelve)* line as indicated by (B) in Fig. 4.3.

*Don't pass (bar twelve)* is *not* the exact opposite of *pass*. How could it be? A bet on pass favors the house. So, the exact opposite would favor the player, and no casino could stay in business for very long offering bets that are favorable to the player. Hence the "bar twelve." What does that mean? Well, the exact opposite of pass would be as follows: You win if the come-out roll loses (produces a 2, 3, or 12) or the point roll loses (a 7 appears before an established point is made). You lose if the come-out-roll wins (7 or 11) or an established point is made on a roll before a 7 appears. Such a bet would give the player 251:244 odds. No way! To bring the bet more in line with the pass bet, no pay-off is made if a 12 appears on the come-out roll; hence the term, "bar twelve." The pay-off is thus as follows: You win if the come-out roll produces a 2 or 3 or if a 7 appears before an established point is made; you break even if a 12 appears on the come-out roll; you lose in all other cases. (Mathematically, it would not make any

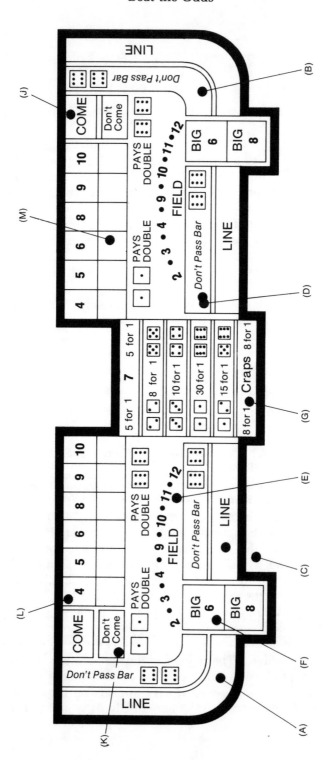

Fig. 4.3 Placement of chips

difference if the 12 were replaced by the 2. They supposedly make that replacement in northern Nevada.)

The don't pass (bar twelve) bet is slightly more favorable to the player than the pass bet, but you may do better still and significantly reduce your expected loss by *taking the odds* on a pass bet or by *laying the odds* on a don't pass bet.

### Taking (Single) Odds on a Pass Bet

*Taking odds* means that you are betting that the shooter will make his point. This bet has to be placed between the come-out roll and the first point roll. You place your stake (which may not exceed your stake on the pass line) behind the pass line [(C) in Fig. 4.3] after a point has been established. You win if you win the pass bet (the shooter makes the point), and you lose if you lose the pass bet (the shooter sevens out).

As innocuous as this bet may appear, it reduces your expected loss dramatically because it pays off at the correct odds:

2:1 if the point is a 4 or a 10
3:2 if the point is a 5 or a 9
6:5 if the point is a 6 or an 8

In conjunction with a pass bet, taking the (single) odds reduces your expected loss to $8.48 per $1000 bet! You can't do much better anywhere, except by laying the odds on a don't pass bet or by playing blackjack (if you are a good blackjack player).

### Laying (Single) Odds on a Don't Pass Bet

*Laying odds* means that you are betting that the shooter won't make his point, that is, that he sevens out before rolling his point. The bet has to be placed between the come-out roll and the first point roll.

You put your chips (in an amount not to exceed your stake on the *don't pass line*) with your chips on the don't pass line by tilting all of them on the edge of the bottom one [see (D) in Fig. 4.3].

As in taking the odds on a pass bet, laying the odds on don't pass pays off at correct odds:

1:2 if the point is a 4 of a 10
2:3 if the point is a 5 or a 9
5:6 if the point is a 6 or 8

In conjunction with a don't pass bet, laying the odds reduces your expected loss to $8.18 per $1000 bet.

When paying off winnings on odds, casinos won't mess with fractions of chips. For example, if you lay out four chips, each of which represents the table minimum, on odds on a don't pass bet (having wagered the same amount on don't pass), and the point happens to be a 5, the casino won't pay off 8/3 chips but only 6/3, or two chips. In general, casinos do not pay off winnings that are not exact multiples of the table minimum. Many of them therefore offer so-called *full odds* (as opposed to *single odds*) by letting you wager a greater stake on odds than on pass or don't pass so as to make an exact pay-off possible. For example, if you have three chips riding on pass and the point is a 5, they will let you augment your odds wager by one chip. In case you win, the pay-off of 3:2 can be made in the exact amount of six chips. On the other hand, if you have two chips riding on don't pass and the point is a 6, they will *not* let you increase your wager on odds to six chips (which is the minimum that will make an exact pay-off of 5:6 possible). You will have to have at least three chips on don't pass for them to do so. Before getting involved with odds wagers, inquire about the local rules for full odds. The dealer will be glad to help you. (The expected losses that we have listed for taking and laying odds are based on an original wager that permits an exact pay-off at single odds.)

The table minimum and table maximum, incidentally, vary from place to place. There are "sawdust joints" (the very opposite of "rug joints") where the table minimum may be as low as two bits, or even a dime. At more reputable casinos, it is not unusual to find a table minimum of one dollar. In Atlantic City, two dollars is rock bottom, and at some tables it is five dollars. The table maximum, again, varies and may be as low as $25 or as high as $500 (on a bet that pays off even money; less on bets that pay off at a higher ratio). Before you start your game, ask the dealer about the table minimum and table maximum (unless it is displayed).

The shooter always has to bet on pass or don't pass. Whether he takes or lays the odds as well is his business. In addition, he may also place any of the other bets which we shall discuss shortly. (Being compelled to bet on pass or don't pass would appear to restrict the shooter's freedom of action. It does. Still most craps players like to roll the dice. It is exhilarating to be the center of the action and, besides, being in control of the dice provides you with just about the only opportunity to cheat!)

For those impatient characters who want to get into the pass or don't pass action as soon as they hit the table and can't wait for the next come-out roll, or those worthies who want their money working all the time to their maximum advantage, there are the *come* and *don't come* wagers. These are, in essence, pass and don't pass wagers, except that they may be initiated at any time. For those who bet on come or don't come, the next roll up is the come-out-roll. You make such a wager by placing your stake into the *come box* or the *don't come box* [see (J) and (K) in Fig. 4.3]. In a way, a come or a don't come bet is a game within a game. A come wager wins if the next roll produces a 7 or 11 and loses if it produces a 2, 3, or 12. If it produces a *come point*, then the dealer moves your stake from the come box to the appropriate *point box*, as indicated by (L) for point 4 in Fig. 4.3. On a don't come wager, the stake is moved from the don't come box to the appropriate don't come point box, as indicated by (M) for point 6 in Fig. 4.3. You may take odds on come and lay odds on don't come just as you do on pass and don't pass.

Obviously, a point on a come bet may be made well after the shooter has rolled his point in the series of rolls during which the come bettor first came aboard, even well into the shooter's next series. (Suppose the come-out roll produces a 4, the next roll a 3, a come bet is placed, and the next rolls produce a 6, 5, 5, and 4. The point has been made on the original come-out roll. The new come-out roll produces a 10, followed by a 6. Only now has the point been made on the come bet.) Don't place too many come or don't come bets when you have other bets riding as well. It is not an easy matter to keep track of them all. Craps is a very fast moving game, and you had better have all your wits about you if you wish to follow the action. Beware of free drinks (or *cave potus liberos* as the ancient Roman dice players would have said)!

Betting "with the dice" *(pass, come)* is often referred to as "betting right" and "betting against the dice" *(don't pass, don't come)* as "betting wrong." These are just technical terms and do not imply any moral or legal judgement. (If it is legally right to bet right, then it is also legally right to bet wrong, and if it is legally wrong to bet wrong, then it is also legally wrong to bet right! It is always morally wrong to bet, no matter what.)

The following wagers are much less advantageous to the player than the ones we have discussed hitherto. Why do people place them? Who knows why people do any of the things they do when immersed in the unreal world of a casino?

### Field

Pay-off:      1:1, except 2:1 on 2 or 12
Odds for winning:      4:5
Expected loss:      $55.56 per $1000 bet

You place your stake as indicated by (E) in Fig. 4.3. You bet that the next roll produces one of the *field numbers*: 2, 3, 4, 9, 10, 11, 12. If the field number is a 2 or 12, the pay-off is 2:1. Otherwise, it is even money.

### Bix Six or Big Eight

>    *Pay-off:*     Even money
>    *Odds for winning:*     5:6
>    *Expected loss:*     $90.90 per $1000 bet.

You place your stake as indicated by (F) in Fig. 4.3 for a bet on big six. You bet that on the rolls to follow (whether come-out rolls or point rolls) a 6 will appear before a 7 in case of the *big six* wager or an 8 will appear before a 7 in case of the *big eight* wager.

### Any Craps

>    *Pay-off:*     7:1
>    *Odds for winning:*     1:8
>    *Expected loss:*     $111.11 per $1000 bet

You place your stakes as indicated by (G) in Fig. 4.3. You bet that the roll immediately following the placing of your bet produces a 2, 3, or 12.

### Seven, Two, Twelve, Three, Eleven

>    *Pay-off:*     4:1 on 7
>                 29:1 on 2 and 12
>                 14:1 on 3 and 11
>    *Odds for Winning:*     1:5 on 7
>                          1:35 on 2 and 12
>                          1:17 on 3 and 11
>    *Expected loss:*     $166.67 per $1000 bet (real sucker bets!)

You toss your chips into the center of the lay-out (see Fig. 4.3) and announce your wager. The stickman will place your chips appropriately. You bet that the designated number (7, 2, 12, 3, or 11) will appear at the next roll.

### Hard Way

>    *Pay-off:*     9:1 on 6 and 8
>                 7:1 on 4 and 10
>    *Odds for winning:*     1:10 on 6 and 8
>                          1:8 on 4 and 10
>    *Expected loss:*     $90.90 on 6 and 8 per $1000 bet
>                       $111.11 on 4 and 10 per $1000 bet

As under (H), you toss your chips into the center of the lay-out and announce your wager. The stickman will place your stake appropriately.

A *hard way* bet on 4, 6, 8, or 10 is a bet that two 2's, two 3's, two 4's, or two 5's, respectively, will be rolled before a 4, 6, 8, or 10 appears in any other combination (1,3; 2,4; 3,5; 4,6, for example) or a 7 appears.

For those who do not wish to conform to the standard bets marked on the table lay-out, many other bets are available: You may wager on the appearance of a specific dice combination on the upcoming roll, such as 1 and 1, 1 and 2, 1 and 3, 2 and 3, etc. (called *hop bets*); you may bet on *any craps and 11*; or you may place a *whirl bet* on 2, 3, 7, 11, 12 and the devil only knows what else. None of these bets offer particularly advantageous odds to the player.

A word on the apparent discrepancy between the pay-off odds as we listed them in the text and the ones marked on the table lay-out (see Figs. 4.2 and 4.3). For example, it says "5 for 1" on a 7 in the center of the table lay-out whereas we stated the pay-off odds to be 4:1. What the casinos mean by "5 for 1" is that if you bet one unit and win, then you get five units, *including your own*, back; in effect, you win only four units. The same is true of all the other cases: "8 for 1" really means 7:1, "10 for 1" means 9:1, "30 for 1" means 29:1, "15 for 1" means 14:1, and "8 for 1" means 7:1. The casino practice of listing the pay-off odds is rather misleading and perhaps intentionally so.

"Friendly craps" is played by the same rules, or similar rules anyway, but the shooter is also the banker and all the other players bet against him. The dice and the bank are passed to the next player when the shooter sevens out.

According to Fielding (Ref. [3], p. 165), there are over 60 different versions of craps. The casinos in Monte Carlo offer the so-called *Idaho variety*. The rules are essentially the same, but the pay-offs on some wagers are different. For example, on a don't pass bet, it is not the 12 on which you break even but the 3. The change makes it a much less attractive bet, raising the expected loss to $41.41 per $1000. A *field bet* pays even money on 2 and 12, as well as the other numbers, raising the expected loss to $111.11 per $1000 bet. These and some other variations make it a much less alluring game than the one offered by American casinos. It should be added that craps is not popular in Europe. Casinos probably offer it only as a concession to American tourists and then proceed to soak them.

## SYSTEMS—SUCH AS THEY ARE

Since craps offers wagers on "even chances," all systems discussed in Chap. 2 that apply to betting on even chances—the *martingale, grande martingale, snowball, paroli,* and *d'Alembert's progres-*

sion—apply to craps as well. Somehow, these systems never caught on among craps players, and there is a very good reason. The possibility of taking (or laying) the odds changes the entire complexion of the game.

From the expected losses that were tabulated in the preceding section under the various bets, it is clear that the prudent gambler should put his wager on pass or don't pass and, by all means, take full advantage of whatever odds the casino has to offer! Don't pay attention to the exhortations of the stickman, who keeps chanting, "Hard way, play the field, eleven, craps," and such (Ref. [5], p. 89).

If full odds are not permitted, bet ten times the table minimum (or multiples thereof) on the pass (come) line or six times the table minimum on don't pass (don't come) so that odds can be correctly paid off. Otherwise, you will not receive the full benefit on the odds wager.

If the casino offers full odds, be sure to put enough on pass (or don't pass) to permit your putting enough chips on odds to make a correct pay-off possible.

If the casino offers *double odds*, then, by all means, avail yourself of the opportunity. (In the long run, you'll lose anyway. You may just as well go down in style.) With double odds, you may wager twice your original stake on odds, taking them or laying them, whatever the case might be. Double odds reduces your expected losses to $6.06 per $1000 bet if you "bet right" and to $5.84 per $1000 bet if you "bet wrong."

Experts recommend that you first bet either on the pass line, taking full advantage of whatever odds are available, or on the don't pass (bar twelve) line, again taking advantage of all the odds you can get. As soon as the point is established, you then place a come bet (if you are betting right) or a don't come bet (if you are betting wrong), and, after that roll, you place another come (or don't come) bet and take (or lay) the odds to the fullest extent in both cases. (If the shooter sevens out, you are back to square one!) If a come bet or don't come bet is paid off, or you lost your stake on it, place another one so that you have two come bets (or don't come bets) working for you at all times.

Your expected losses are (marginally) smaller if you bet wrong rather than right. Still, most people prefer to bet right. Superstition?

In Table 4.1, we have taken a sequence of rolls—actually generated by two dice under battlefield conditions—and traced back what would have happened had we played the pass line, taking single or double odds and having two come bets working for us all the time, and by contrast, betting on don't pass, laying single or double odds, and having two don't come bets working all the time. We used chips representing $30 each (as represented by the full circles sitting on

**Table 4.1   Taking and laying of odds**

| Points | Pass single odds | Pass double odds | Don't Pass single odds | Don't Pass double odds | Come single odds | Come double odds | Don't Come single odds | Don't Come double odds | Come single odds | Come double odds | Don't Come single odds | Don't Come double odds | Come single odds | Come double odds | Don't Come single odds | Don't Come double odds |
|---|---|---|---|---|---|---|---|---|---|---|---|---|---|---|---|---|
| 8 | • | • | • | • | | | | | | | | | | | | |
| 6 | • | • | • | • | • | • | • | • | | | | | | | | |
| 8 | +66 | +102 | −60 | −90 | • | • | • | • | • | • | • | • | | | | |
| 12 | | | | | | | | | • | • | • | • | | | | |
| 5 | | | | | | | | | | | | | | | | |
| 5 | | | | | | | | | | | | | | | | |
| 10 | | | | | | | | | | | | | | | | |
| 8 | | | | | | | | | +66 | +102 | −60 | −90 | | | | |
| 2 | | | | | | | | | | | | | −30 | −30 | +30 | +30 |
| 5 | | | | | | | | | • | • | • | • | | | | |
| 10 | | | | | | | | | | | | | | | | |
| 6 | | | | | +66 | +102 | −60 | −90 | | | | | | | | |
| 6 | | | | | | | | | | | | | • | • | • | • |
| 4 | | | | | | | | | | | | | • | • | • | • |
| 8 | | | | | | | | | | | | | | | | |
| 7 | | | | | | | | | −60 | −90 | +50 | +70 | −60 | −90 | +55 | +80 |

horizontal lines) so that odds could be paid off exactly in either case. The horizontal lines represent betting times between rolls, and the fat horizontal bars represent the end of a particular series when some bets are either paid off or collected. The numbers sitting on the fat bars

represent pay-offs when + and losses when −. The results for this
particular case are as follows:

Playing the pass line taking single odds ............. $48 WIN
Playing the pass line taking double odds ............. $96 WIN
Playing the don't pass line taking single odds ....... $45 LOSS
Playing the don't pass line taking double odds ...... $90 LOSS

In this case, betting right turned out to be more advantageous.
(The fact that the losses on the don't pass line are not the exact
opposite of the winnings on the pass line is due to the fact that the
odds for winning a pass bet are not exactly the same as the odds for
winning a don't pass bet. Remember that the odds for winning a pass
bet are 244:251, whereas the odds for winning a don't pass (bar twelve)
bet are 217:223.

## THE COMPUTER PROGRAM

The computer program in Fig. 4.4, which produced the printout
in Fig. 4.1, simulates a craps game in which only the operator may
place bets. To provide for some variety, we have included some
fictional characters to serve as other players. All they do, however, is
shoot the dice whenever it is their turn. The operator takes his turn at
shooting with Friendly-Friendly, Paddy the Cork, Chicago Sam, Ciel
Magfil, and Little Jim. (We were tempted to let Little Jim, whom we
perceive as intimidating and somewhat less than honest, shoot with
invisible dice, but, in the end, decided against it. Nothing is so
frustrating as watching a computer hum and click without producing
any output!)
We let the shooter complete his series of rolls in all cases where
a wager has been placed on pass, don't pass, field, any craps, 7, 11, 3,
2, 12 but terminate the rolls on wagers on big six, big eight, and hard
way as soon as a decision has been reached. It would not be necessary
to play out the series in the case of field, any craps, 7, 11, 3, 2, 12
either since a decision is reached on the roll that follows the placing
of the bet, but completing the series adds realism to the simulation. To
do likewise for big six, big eight, and hard way would have complicated the program without serving a useful purpose. (Remember that
bets on big six, big eight, and hard way may not necessarily be decided
within one series of rolls. For example, if you bet on big six and the
subsequent rolls produce the series 5, 3, 11, 8, 3, 4, 5, 10, 6, then you
win two rolls into the second series, the first series ending when the
point 5 is made on the seventh roll.)
The only "new" technique that has not been used in the preceding programs is the simulation of rolls with two dice, and that is not

really new. It is just another application of the sub-string capability of our computer. Since the rolls have to be accessible from many points in our program, we put them into a subroutine:

```
2350   I=INT(RND(1)*6)+1
2360   J=INT(RND(1)*6)+1
2370   PRINT TAB(25);F$[1+3*(I-1),3*I];TAB(34);F$[1+3*(J-1),3*J]
2380   PRINT TAB(25);S$[1+3*(I-1),3*I];TAB(34);S$[1+3*(J-1),3*J]
2390   PRINT TAB(25);T$[1+3*(I-1),3*I];TAB(34);T$[1+3*(J-1),3*J]
2400   RETURN
```

where the First, Second, and Third string variables—F$, S$, and T$—are defined in the following lines:

```
50   F$="    *    *    *  **  ****"
60   S$=" *           *        *       "
70   T$="       *   **  **  ****"
```

As in all our programs, we do not require an input in a complicated numerical code when asking the operator to place his bet. We ask for an input in plain English. The computer then proceeds to encode the input from line 710 by means of a "recognition loop" similar to the ones we used in trente-et-quarante (lines 330–350 in Fig. 1.5) and roulette (lines 790–830 and 1060–1080 in Fig. 2.14). If we look at the 12 possible inputs,

| P | A | S S |   |   |         |
|---|---|-----|---|---|---------|
| D | O | N ' | T |   | P   A   S   S |
| F | I | E L | D |   |         |
| B | I | G   | S | I | X       |
| B | I | G   | E | I | G   H   T |
| A | N | Y   | C | R | A   P   S |
| 7 |   |     |   |   |         |
| 1 | 1 |     |   |   |         |
| 2 |   |     |   |   |         |
| 1 | 2 |     |   |   |         |
| 3 |   |     |   |   |         |
| H | A | R D |   |   | W   A   Y |

we recognize that the computer will have to scan the first, second, and fifth position to obtain 12 distinguishable readings. (The fifth position is required to separate the big six from the big eight.) We introduce the "test" string,

```
100   B$="PDFBBA71213HAOIIIN 1 2 A TDSEC
```

and encode the input from 710 by means of the loop,

```
760   FOR L=1 TO 12
770   IF C$[1,1] <> B$[L,L] THEN 810
780   IF C$[2,2] <> B$[12+L,12+L] THEN 810
790   IF C$[5,5] <> B$[24+L,24+L] THEN 810
800   GOTO 830
810   NEXT L
820   IF L=13 THEN 700
```

In line 820 we took out insurance against a faulty input in line 710 that would otherwise lead to a pass, unless the faulty input agrees with one of the permissible inputs in the first, second, or fifth positions.

The shooter may *seven out* at various points of the program. Hence, we put the "sevening out" into a subroutine and, for efficiency's sake, made it part of the "point-rolls subroutine," which also has to be accessible from many points:

```
2220   PRINT "NOW COME THE POINT ROLLS";
2230   GOSUB 2350
2240   PRINT
2250   IF I+J=P THEN 2320
2260   IF I+J <> 7 THEN 2230
2270   PRINT P$[1+17*(S-1),17*S]" SEVENS OUT AND "
2280   PRINT P$[1+17*S,17*(S+1)]" IS THE NEXT SHOOTER."
2290   PRINT LIN(1)
2300   S=S+1-6*INT(S/6)
2310   GOTO 2340
2320   PRINT "POINT !"
2330   PRINT
2340   RETURN
```

Note how the "sevening out subroutine" may be accessed independently of the "point rolls" subroutine by means of GOSUB 2270. (To make the program easier to read, we programmed separate point rolls routines for pass and don't pass bets; see lines 1080–1120 and 1420–1500.) Note that the point rolls and "winding up the series" routines in lines 1610 to 1640 are also accessed from lines 1810, 1830, 1870, 1890, 1930, 1950, 1990, and 2010.

The bookkeeping part of our program has to be accessed after a decision has fallen on a wager from eight different points of our program. Again, we put it into a subroutine:

```
2410   IF B>0 THEN 2450
2420   IF B<0 THEN 2470
2430   PRINT "YOU BROKE EVEN ";
2440   GOTO 2480
2450   PRINT "YOU WON $ "B;
2460   GOTO 2480
2470   PRINT "YOU LOST $ "-B;
2480   M=M+B
2490   PRINT "AND YOU NOW HAVE $ "M
2500   PRINT LIN(1)
2510   IF M <= 0 THEN 2530
2520   RETURN
```

We let the operator take (or lay) single odds on the point roll in conjunction with a pass (don't pass) wager. We assume that the table minimum is \$1 and the pay-off is in whole numbers only, using the INT-function for that purpose (see lines 1150 and 1470). Since an odds wager pays off at correct odds and since the probabilities for rolling P = 4, 5, 6, 8, 9, or 10 are 3/36, 4/36, 5/36, 5/36, 4/36, and 3/36, respectively, and since the probability for rolling P = 7 is 6/36, we may list the pay-off odds for taking the odds as

$$6 \text{ to } (P-1) \text{ MIN } (13-P)$$

and for laying the odds as

$$(P-1) \text{ MIN } (13-P) \text{ to } 6$$

if we also observe that

$$(P-1) \text{ MIN } (13-P) = \begin{cases} 3 \text{ for } P = 4,10 \\ 4 \text{ for } P = 5,9 \\ 5 \text{ for } P = 6,8 \end{cases}$$

(see also lines 1030 and 1370).

In lines 1000 and 1340, we make sure that the operator is not asked if he wants to take (lay) odds if he does not have enough money left to cover such a wager.

Finally, let us say a word about the strange entries in lines 1660, 1910, and 1970.

Since the pay-off is the same whether you bet on the big six or the big eight (1:1), the same for a wager on 3 or 11 (14:1), and the same for a wager on 2 or 12 (29:1), it was expedient to treat wagers with the same pay-off jointly. Towards that end, we note that the recognition loop 760–810 assigns the code numbers L = 4 to the big six; L = 5 to the big eight; L = 8, 11 to bets on 11, 3; and L = 9, 10 to bets on 2, 12. Since

$$5*\text{SGN}(P-6) + 4*\text{SGN}(8-P) = \begin{cases} 4 \text{ if and only if } P = 6 \\ 5 \text{ if and only if } P = 8 \end{cases}$$

$$11*\text{SGN}(11-P) + 8*\text{SGN}(P-3) = \begin{cases} 8 \text{ if and only if } P = 11 \\ 11 \text{ if and only if } P = 3 \end{cases}$$

$$9*\text{SGN}(12-P) + 10*\text{SGN}(P-2) = \begin{cases} 9 \text{ if and only if } P = 2 \\ 10 \text{ if and only if } P = 12 \end{cases}$$

we can say that the expression on the left is equal to L if and only if the operator won his bet.

```
10    PRINT TAB(27)"*********"
20    PRINT TAB(27)"* CRAPS *"
30    PRINT TAB(27)"*********"
40    DIM F$[18],S$[18],T$[18],P$[204],B$[36],N$[20],R[6],C$[10]
50    F$="  *   *   *  **  ****"
60    S$=" *     *      *       "
70    T$="   *      *  ** **  ****"
80    P$[1,43]="FRIENDLY-FRIENDLYPADDY THE CORK   CIEL MAGF"
90    P$[44,85]="IL       CHICAGO SAM       LITTLE JIM       "
100    B$="PDFBBA71213HAOIIIN 1 2 A TDSEC       "
110   PRINT LIN(1)
120   PRINT "WHEN ASKED TO 'NAME YOUR BET', ENTER ONE OF THE ";
130   PRINT "FOLLOWING:"
140   PRINT
150   PRINT TAB(9)"PASS"TAB(20)"(PAY-OFF 1:1)"
160   PRINT TAB(9)"DON'T PASS"TAB(20)"(PAY-OFF 1:1)"
170   PRINT TAB(9)"FIELD"TAB(20)"(PAY-OFF 1:1, 2:1 ON 2,12)"
180   PRINT TAB(9)"BIG SIX"TAB(20)"(PAY-OFF 1:1)"
190   PRINT TAB(9)"BIG EIGHT"TAB(20)"(PAY-OFF 1:1)"
200   PRINT TAB(9)"ANY CRAPS"TAB(20)"(PAY-OFF 7:1)"
210   PRINT TAB(9)"7"TAB(20)"(PAY-OFF 4:1)"
220   PRINT TAB(9)"11"TAB(20)"(PAY-OFF 14:1)"
230   PRINT TAB(9)"3"TAB(20)"(PAY-OFF 14:1)"
240   PRINT TAB(9)"2"TAB(20)"(PAY-OFF 29:1)"
250   PRINT TAB(9)"12"TAB(20)"(PAY-OFF 29:1)"
260   PRINT TAB(9)"HARD WAY"TAB(20)"(PAY-OFF 9:1 ON 6,8 ";
270   PRINT "AND 7:1 ON 4,10)"
280   PRINT LIN(1)
290   PRINT TAB(7)"*** SHOOTER MUST WAGER ON PASS OR DON'T ";
300   PRINT "PASS ***"
310   PRINT LIN(1)
320   PRINT "PLEASE IDENTIFY YOURSELF ";
330   INPUT N$
340   PRINT
345   REM LOOP 350 TO 370 LOOKS FOR THE BLANK SPACE BETWEEN
346   REM FIRST NAME AND SURNAME OF THE OPERATOR, AS ENTERED
347   REM IN LINE 330
350   FOR Z=1 TO 20
360   IF N$[Z,Z]=" " THEN 380
370   NEXT Z
380   P$[86,91]="SLICK "
390   P$[92,102]=N$[1,Z-1]
395   REM P$, AS DEFINED IN 80,90,380,390, CONTAINS THE
396   REM NAMES OF ALL PLAYERS, INCLUDING THE OEPERATOR'S
400   P$[103,204]=P$
405   REM THE DEFINITION OF P$ IS EXTENDED PERIODICALLY IN
406   REM LINE 400 BECAUSE, WHEN THE OPERATOR 'SEVENS OUT'
407   REM IN LINE 2270, THE ARGUMENT OF P$ IN LINE 2280
408   REM BECOMES (103,119) BEFORE S IS SET BACK MODULO 6
409   REM AND AUGMENTED BY 1 IN LINE 2300
410   PRINT "HELLO "P$[86,91+Z]". MEET THE OTHER CHARACTERS WHO"
420   PRINT "CROWD AROUND THE CRAP TABLE: FRIENDLY-FRIENDLY, ";
430   PRINT "PADDY"
440   PRINT "THE CORK, CIEL MAGFIL, CHICAGO SAM, AND LITTLE JIM."
450   PRINT
460   PRINT "HOW MUCH MONEY DO YOU HAVE ";
470   INPUT M
480   PRINT "YOU WON'T HAVE IT FOR LONG !"
490   PRINT LIN(1)
500   PRINT TAB(9)"*** FIRST HIGH ROLL GETS TO SHOOT FIRST ***"
510   FOR K=1 TO 6
520   PRINT
```

*Fig. 4.4 Master program "Craps"*

```
530    PRINT P$[1+17*(K-1),17*K]" ROLLS ";
540    GOSUB 2350
550    R[K]=I+J
560    NEXT K
565    REM LOOP 570 TO 620 PICKS OUT THE FIRST HIGH ROLL
570    FOR K=1 TO 6
580    FOR L=K+1 TO 6
590    IF R[K]<R[L] THEN 620
600    NEXT L
610    GOTO 630
620    NEXT K
630    S=K
640    PRINT LIN(1)
650    PRINT TAB(11)"*** "P$[1+17*(S-1),17*S]" IS COMING OUT ***"
660    PRINT LIN(1)
670    IF P$[1+17*(S-1),17*S] <> P$[86,102] THEN 700
675    REM IF THE OPERATOR IS THE SHOOTER, HE HAS TO WAGER
676    REM ON 'PASS' OR 'DON'T PASS'. HENCE LINES 680,690
680    PRINT "DO YOU WANT TO BET ON 'PASS' OR 'DON'T PASS' ";
690    GOTO 710
700    PRINT P$[86,91+Z]", NAME YOUR BET ";
710    INPUT C$
720    PRINT
730    PRINT "HOW MUCH DO YOU WANT TO BET ";
740    INPUT B
750    IF M-B<0 THEN 730
755    REM LOOP 750 TO 790 CHECKS WHICH OF THE 12 BETS
756    REM LISTED IN LINES 150 TO 270 HAS BEEN ENTERED
757    REM IN LINE 710 AS C$
760    FOR L=1 TO 12
770    IF C$[1,1] <> B$[L,L] THEN 810
780    IF C$[2,2] <> B$[12+L,12+L] THEN 810
790    IF C$[5,5] <> B$[24+L,24+L] THEN 810
800    GOTO 830
810    NEXT L
820    IF L=13 THEN 700
825    REM FOR L=12 (A 'HARD WAY' BET), ADDITIONAL INFOR-
826    REM MATION HAS TO  BE ENTERED IN LINE 850
830    IF L <> 12 THEN 860
840    PRINT "ON 6, 8, 4, OR 10 ";
850    INPUT Q
860    PRINT LIN(1)
870    PRINT "ON THE COME-OUT ROLL, "P$[1+17*(S-1),17*S]" ROLLS "
880    PRINT
890    GOSUB 2350
900    PRINT
905    REM IN LINES 910 TO 930, CONTROL IS PASSED TO THAT
906    REM PORTION OF THE PROGRAM THAT DEALS WITH THE WAGER
907    REM THAT HAS BEEN ENTERED IN LINE 710
910    IF L=11 THEN 1900
920    IF L=12 THEN 2020
930    GOTO L OF 940,1250,1550,1650,1650,1780,1840,1900,1960,1960
935    REM ONE DIE ROLLS I POINTS, THE OTHER ONE J POINTS
936    REM FOR A TOTAL OF P=I+J POINTS
939    REM LINES 940 TO 1240 DEAL WITH 'PASS' BETS
940    P=I+J
950    IF P=7 OR P=11 THEN 1230
960    IF P <> 2 AND P <> 3 AND P <> 12 THEN 990
970    B=-B
980    GOTO 1230
990    PRINT "POINT "P
995    REM IF THE OPERATOR CANNOT COVER AN 'ODDS' WAGER,
```

Fig. 4.4 Master program "Craps" (cont'd)

```
996    REM HE ISN'T EVEN ASKED IF HE WANTS TO. SEE LINE 1000
1000   IF M-2*B<0 THEN 1060
1010   PRINT
1020   PRINT "DO YOU WANT TO TAKE SINGLE ODDS AT A PAYOFF OF ";
1030   PRINT "6 TO "(P-1) MIN (13-P);
1040   INPUT C$
1050   GOTO 1070
1060   C$="NO"
1070   PRINT
1080   PRINT "NOW COME THE POINT ROLLS";
1090   GOSUB 2350
1100   PRINT
1110   IF I+J <> P THEN 1170
1120   PRINT "POINT !"
1130   PRINT
1140   IF C$[1,1]="N" THEN 1230
1150   B=INT(B+6*B/((P-1) MIN (13-P)))
1160   GOTO 1230
1170   IF I+J <> 7 THEN 1090
1180   IF C$[1,1]="N" THEN 1210
1190   B=-2*B
1200   GOTO 1220
1210   B=-B
1220   GOSUB 2270
1230   GOSUB 2410
1240   GOTO 650
1245   REM LINES 1250 TO 1540 DEAL WITH 'DON'T PASS' BETS
1250   P=I+J
1260   IF P=2 OR P=3 THEN 1230
1270   IF P <> 12 THEN 1300
1280   B=0
1290   GOTO 1230
1300   IF P <> 7 AND P <> 11 THEN 1330
1310   B=-B
1320   GOTO 1230
1330   PRINT "POINT "P
1335   REM IF THE OPERATOR CANNOT COVER AN 'ODDS' WAGER,
1336   REM HE ISN'T EVEN ASKED IF HE WANTS TO. SEE LINE 1340
1340   IF M-2*B<0 THEN 1400
1350   PRINT
1360   PRINT "DO YOU WANT TO LAY SINGLE ODDS AT A PAY-OFF OF ";
1370   PRINT (P-1) MIN (13-P)" TO 6";
1380   INPUT C$
1390   GOTO 1410
1400   C$="NO"
1410   PRINT
1420   PRINT "NOW COME THE POINT ROLLS";
1430   GOSUB 2350
1440   PRINT
1450   IF I+J <> 7 THEN 1490
1460   IF C$[1,1]="N" THEN 1220
1470   B=INT(B+((P-1) MIN (13-P))*B/6)
1480   GOTO 1220
1490   IF I+J <> P THEN 1430
1500   PRINT "POINT !"
1510   PRINT
1520   IF C$[1,1]="N" THEN 1310
1530   B=-2*B
1540   GOTO 1230
1545   REM LINES 1550 TO 1640 DEAL WITH 'FIELD' BETS
1550   P=I+J
1560   IF P<5 OR P>8 THEN 1590
1570   B=-B
```

*Fig. 4.4 Master program "Craps" (cont'd)*

```
1580    GOTO 1610
1590    IF P <> 2 AND P <> 12 THEN 1610
1600    B=2*B
1604    REM LINES 1610 TO 1640 PLAY OUT A SERIES AFTER A DECISION
1605    REM HAS BEEN REACHED (EXCEPT FOR 'BIG SIX', BIG EIGHT',
1606    REM AND 'HARD WAY' BETS) IF THE SHOOTER HAS NEITHER
1607    REM 'SEVENED OUT' NOR MADE HIS POINT. THIS GROUP OF IN-
1608    REM STRUCTIONS IS ALSO ACCESSED FROM LINES 1810,1830,
1609    REM 1870,1890,1930,1950,1990,2010 FOR THE SAME PURPOSE
1610    GOSUB 2410
1620    IF P<4 OR P=7 OR P>10 THEN 650
1630    GOSUB 2200
1640    GOTO 650
1645    REM LINES 1650 TO 1770 DEAL WITH 'BIG SIX' AND
1646    REM 'BIG EIGHT' BETS. NOTE IN 1660 AND 1730 THAT
1647    REM L=4 FOR 'BIG SIX' AND L=5 FOR 'BIG EIGHT'
1650    P=I+J
1660    IF 5*SGN(P-6)+4*SGN(8-P) <> L THEN 1680
1670    GOTO 1230
1680    IF P=7 THEN 1760
1690    PRINT "KEEP TRYING ";
1700    GOSUB 2350
1710    PRINT
1720    P=I+J
1730    IF 5*SGN(P-6)+4*SGN(8-P)=L THEN 1230
1740    IF I+J <> 7 THEN 1700
1750    GOSUB 2270
1760    B=-B
1770    GOTO 1230
1775    REM LINES 1780 TO 1830 DEAL WITH A WAGER ON 'ANY CRAPS'
1780    P=I+J
1790    IF P <> 2 AND P <> 3 AND P <> 12 THEN 1820
1800    B=7*B
1810    GOTO 1610
1820    B=-B
1830    GOTO 1610
1835    REM LINES 1840 TO 1890 TAKE CARE OF A BET ON 7
1840    P=I+J
1850    IF P <> 7 THEN 1880
1860    B=4*B
1870    GOTO 1610
1880    B=-B
1890    GOTO 1610
1895    REM LINES 1900 TO 1950 TAKE CARE OF BETS ON 3 AND ON 11.
1896    REM NOTE THAT L=11 FOR P=3 AND L=8 FOR P=11
1900    P=I+J
1910    IF 11*SGN(11-P)+8*SGN(P-3)=L THEN 1940
1920    B=-B
1930    GOTO 1610
1940    B=14*B
1950    GOTO 1610
1955    REM LINES 1960 TO 2010 TAKE CARE OF BETS ON 2 AND ON 12.
1956    REM NOTE THAT L=9 FOR P=2 AND L=10 FOR P=12
1960    P=I+J
1970    IF 9*SGN(12-P)+10*SGN(P-2)=L THEN 2000
1980    B=-B
1990    GOTO 1610
2000    B=29*B
2010    GOTO 1610
2015    REM LINES 2020 TO 2190 TAKE CARE OF 'HARD WAY' BETS.
2016    REM TO WIN A BET ON Q THE HARD WAY, IT IS NECESSARY
2017    REM AND SUFFICIENT THAT P=Q AND I=Q/2
2020    P=I+J
```

Fig. 4.4 Master program "Craps" (cont'd)

```
2030    IF P=Q AND I=Q/2 THEN 2150
2040    IF P=Q OR P=7 THEN 2130
2050    PRINT "LET'S TRY AGAIN ";
2060    GOSUB 2350
2070    PRINT
2080    IF I+J=Q AND I=Q/2 THEN 2150
2090    IF I+J=Q THEN 2130
2100    IF I+J=7 THEN 2120
2110    GOTO 2060
2120    GOSUB 2270
2130    B=-B
2140    GOTO 2190
2150    IF Q=4 OR Q=10 THEN 2180
2160    B=9*B
2170    GOTO 2190
2180    B=7*B
2190    GOTO 1230
2195    REM THE SUB-ROUTINE 2200 TO 2340 TAKES CARE OF POINT
2196    REM ROLLS, INCLUDING THE POSSIBILITY OF THE SHOOTER
2197    REM 'SEVENING OUT'
2200    PRINT "POINT "P
2210    PRINT
2220    PRINT "NOW COME THE POINT ROLLS";
2230    GOSUB 2350
2240    PRINT
2250    IF I+J=P THEN 2320
2260    IF I+J <> 7 THEN 2230
2265    REM THE SUB-ROUTINE 2270 TO 2310,2340  TAKES CARE OF THE
2266    REM 'SEVENING OUT' AND CAN BE ACCESSED INDEPENDENTLY OF
2267    REM THE POINT ROLLS SUBROUTINE OF WHICH IT IS A PART
2270    PRINT P$[1+17*(S-1),17*S]" SEVENS OUT AND "
2280    PRINT P$[1+17*S,17*(S+1)]" IS THE NEXT SHOOTER."
2290    PRINT LIN(1)
2295    REM IN 2300, S IS SET BACK MODULO 6 AND AUGMENTED BY 1
2296    REM SINCE PLAYER NUMBER 7 IS THE SAME AS PLAYER NUMBER 1
2300    S=S+1-6*INT(S/6)
2310    GOTO 2340
2320    PRINT "POINT !"
2330    PRINT
2340    RETURN
2345    REM THE SUB-ROUTINE 2350 TO 2400 PRODUCES THE ROLL OF
2346    REM TWO DICE
2350    I=INT(RND(1)*6)+1
2360    J=INT(RND(1)*6)+1
2370    PRINT TAB(25);F$[1+3*(I-1),3*I];TAB(34);F$[1+3*(J-1),3*J]
2380    PRINT TAB(25);S$[1+3*(I-1),3*I];TAB(34);S$[1+3*(J-1),3*J]
2390    PRINT TAB(25);T$[1+3*(I-1),3*I];TAB(34);T$[1+3*(J-1),3*J]
2400    RETURN
2405    REM THE SUB-ROUTINE 2410 TO 2520 DOES THE BOOKKEEPING
2410    IF B>0 THEN 2450
2420    IF B<0 THEN 2470
2430    PRINT "YOU BROKE EVEN ";
2440    GOTO 2480
2450    PRINT "YOU WON $ "B;
2460    GOTO 2480
2470    PRINT "YOU LOST $ "-B;
2480    M=M+B
2490    PRINT "AND YOU NOW HAVE $ "M
2500    PRINT LIN(1)
2510    IF M <= 0 THEN 2530
2520    RETURN
2530    PRINT "YOU GOT CLEANED OUT REAL GOOD !"
2540    END
```

Fig. 4.4 Master program "Craps" (cont'd)

## SOME MODIFICATIONS OF THE COMPUTER PROGRAM

Trivial modifications to provide opportunities for placing other wagers such as *hop bets, whirl bets*, and the like, are left to the reader. We also leave it to the reader to set a table minimum and a table maximum and make provisions for the computer to reject bets below the minimum and above the maximum.

If stripped of all extraneous features, the program in Fig. 4.4 may be considerably shortened and simplified. Rather than undertake such a destructive task, let us refer the reader to a program that is simple to begin with, namely the one by Spencer (Ref. [11], p. 92ff). In essence, Spencer's program generates a series of rolls, from the come-out roll to final disposition, and informs the reader whether he won or lost on the come-out roll or the last point roll.

Without destroying the versatility of our program, a number of changes will speed up the play. For example, considerable time is saved if the results of the rolls are printed in the left margin rather than in the middle of the page:

Omit TAB(25) in lines 2370, 2380, and 2390
Omit the semicolon at the end of lines 530, 1080, 1420, 1690, 2050, and 2220

You may cut down on the playing time still further by replacing the "graphic" representation of the dice by numbers, as follows:

Omit lines 50, 60, and 70
2370 PRINT I; J
Omit lines 2380 and 2390

For further simplification, we may cut out the superfluous point rolls when a bet has been placed on field, any craps, 7, 11, 3, 2, or 12, as follows:

Omit line 1630

If we want to permit the taking (laying) of *double odds* (remember that taking—laying—double odds means that you wager twice your original stake on odds) on the point rolls of a pass (don't pass) bet, we may proceed as follows:

```
1000   IF M−3*B<0 THEN 1060
1020   PRINT "DO YOU WANT TO TAKE DOUBLE ODDS AT A PAY-OFF";
1150   B = INT(B+12*B/((P−1) MIN(13−P)))
1190   B = −3*B
1340   IF M−3*B<0 THEN 1400
1360   PRINT "DO YOU WANT TO LAY DOUBLE ODDS AT A PAY-OFF";
1470   B = INT(B+((P−1) MIN(13−P))*B/3)
1530   B = −3*B
```

(Note that when you lose, you lose three times your original wager:

the original wager itself and twice the original wager that has been placed on odds.)

   With all these changes incorporated into our program, we ran it for a while, all the time betting on don't pass and laying double odds. To derive full benefit from the double odds wager, we put integral multiples of three times the table minimum of $1 on don't pass. Here is what happened (note how it swings back and forth):

```
HOW MUCH MONEY DO YOU HAVE ?100
YOU WON'T HAVE IT FOR LONG !


              *** CIEL MAGFIL       IS COMING OUT ***

SLICK HANS , NAME YOUR BET ?DON'T PASS

HOW MUCH DO YOU WANT TO BET ?3


ON THE COME-OUT ROLL, CIEL MAGFIL      ROLLS

  3     4

YOU LOST $  3    AND YOU NOW HAVE $  97


              *** CIEL MAGFIL       IS COMING OUT ***

SLICK HANS , NAME YOUR BET ?DON'T PASS

HOW MUCH DO YOU WANT TO BET ?6


ON THE COME-OUT ROLL, CIEL MAGFIL      ROLLS

  6     6

YOU BROKE EVEN AND YOU NOW HAVE $  97


              *** CIEL MAGFIL       IS COMING OUT ***

SLICK HANS , NAME YOUR BET ?DON'T PASS

HOW MUCH DO YOU WANT TO BET ?6


ON THE COME-OUT ROLL, CIEL MAGFIL      ROLLS

  2     5

YOU LOST $  6    AND YOU NOW HAVE $  91


              *** CIEL MAGFIL       IS COMING OUT ***
```

SLICK HANS , NAME YOUR BET ?DON'T PASS

HOW MUCH DO YOU WANT TO BET ?6


ON THE COME-OUT ROLL, CIEL MAGFIL          ROLLS

   4      3
YOU LOST $   6     AND YOU NOW HAVE $   85


            *** CIEL MAGFIL          IS COMING OUT ***


SLICK HANS , NAME YOUR BET ?DON'T PASS

HOW MUCH DO YOU WANT TO BET ?6


ON THE COME-OUT ROLL, CIEL MAGFIL          ROLLS

   2      1

YOU WON $   6     AND YOU NOW HAVE $   91


            *** CIEL MAGFIL          IS COMING OUT ***


SLICK HANS , NAME YOUR BET ?DON'T PASS

HOW MUCH DO YOU WANT TO BET ?12


ON THE COME-OUT ROLL, CIEL MAGFIL          ROLLS

   2      6

POINT  8

DO YOU WANT TO LAY DOUBLE ODDS AT A PAY-OFF 5      TO 6?YES

NOW COME THE POINT ROLLS
   6      5

   4      6

   3      3

   6      5

   5      2

CIEL MAGFIL          SEVENS OUT AND
CHICAGO SAM          IS THE NEXT SHOOTER.


YOU WON $   32    AND YOU NOW HAVE $   123


            *** CHICAGO SAM          IS COMING OUT ***

```
SLICK HANS , NAME YOUR BET ?DON'T PASS

HOW MUCH DO YOU WANT TO BET ?12


ON THE COME-OUT ROLL, CHICAGO SAM          ROLLS

   2      3
POINT  5

DO YOU WANT TO LAY DOUBLE ODDS AT A PAY-OFF 4     TO 6?YES

NOW COME THE POINT ROLLS
   6      6

   6      1
CHICAGO SAM           SEVENS OUT AND
LITTLE JIM            IS THE NEXT SHOOTER.


YOU WON $   28    AND YOU NOW HAVE $   151

           *** LITTLE JIM          IS COMING OUT ***


SLICK HANS , NAME YOUR BET ?DON'T PASS

HOW MUCH DO YOU WANT TO BET ?36


ON THE COME-OUT ROLL, LITTLE JIM          ROLLS

   5      4
POINT  9

DO YOU WANT TO LAY DOUBLE ODDS AT A PAY-OFF 4     TO 6?YES

NOW COME THE POINT ROLLS
   6      5

   3      3

   6      4

   5      5

   6      5

   4      3
LITTLE JIM            SEVENS OUT AND
SLICK HANS            IS THE NEXT SHOOTER.


YOU WON $   84    AND YOU NOW HAVE $   235

           *** SLICK HANS          IS COMING OUT ***
```

DO YOU WANT TO BET ON 'PASS' OR 'DON'T PASS' ?DON'T PASS

HOW MUCH DO YOU WANT TO BET ?24


ON THE COME-OUT ROLL, SLICK HANS          ROLLS

  2     2

POINT   4

DO YOU WANT TO LAY DOUBLE ODDS AT A PAY-OFF 3     TO 6?YES

NOW COME THE POINT ROLLS
  1     3

POINT !

YOU LOST $  72   AND YOU NOW HAVE $  163


            *** SLICK HANS          IS COMING OUT ***


DO YOU WANT TO BET ON 'PASS' OR 'DON'T PASS' ?DON'T PASS

HOW MUCH DO YOU WANT TO BET ?24


ON THE COME-OUT ROLL, SLICK HANS          ROLLS

  6     5

YOU LOST $  24   AND YOU NOW HAVE $  139


            *** SLICK HANS          IS COMING OUT ***

DO YOU WANT TO BET ON 'PASS' OR 'DON'T PASS' ?DON'T PASS

HOW MUCH DO YOU WANT TO BET ?12


ON THE COME-OUT ROLL, SLICK HANS          ROLLS

  4     2

POINT   6

DO YOU WANT TO LAY DOUBLE ODDS AT A PAY-OFF 5     TO 6?YES

NOW COME THE POINT ROLLS
  6     6

  2     5

SLICK HANS          SEVENS OUT AND
FRIENDLY-FRIENDLY IS THE NEXT SHOOTER.


YOU WON $  32   AND YOU NOW HAVE $  171

```
                  *** FRIENDLY-FRIENDLY IS COMING OUT ***

SLICK HANS , NAME YOUR BET ?DON'T PASS

HOW MUCH DO YOU WANT TO BET ?36


ON THE COME-OUT ROLL, FRIENDLY-FRIENDLY ROLLS

  6     3

POINT 9

DO YOU WANT TO LAY DOUBLE ODDS AT A PAY-OFF 4     TO 6?YES

NOW COME THE POINT ROLLS
  1     4

  3     3

  2     2

  4     4

  5     4

POINT !

YOU LOST $  108  AND YOU NOW HAVE $  63


                  *** FRIENDLY-FRIENDLY IS COMING OUT ***


SLICK HANS , NAME YOUR BET ?DON'T PASS

HOW MUCH DO YOU WANT TO BET ?12


ON THE COME-OUT ROLL, FRIENDLY-FRIENDLY ROLLS

  3     6

POINT 9

DO YOU WANT TO LAY DOUBLE ODDS AT A PAY-OFF 4     TO 6?YES

NOW COME THE POINT ROLLS
  3     2

  3     5

  1     6


FRIENDLY-FRIENDLY SEVENS OUT AND
PADDY THE CORK    IS THE NEXT SHOOTER.

YOU WON $  28   AND YOU NOW HAVE $  91
```

```
        *** PADDY THE CORK    IS COMING OUT ***


SLICK HANS , NAME YOUR BET ?DON'T PASS

HOW MUCH DO YOU WANT TO BET ?12


ON THE COME-OUT ROLL, PADDY THE CORK     ROLLS

   1     5

POINT  6

DO YOU WANT TO LAY DOUBLE ODDS AT A PAY-OFF 5     TO 6?YES

NOW COME THE POINT ROLLS
   6     2

   5     2

PADDY THE CORK     SEVENS OUT AND
CIEL MAGFIL        IS THE NEXT SHOOTER.


YOU WON $  32   AND YOU NOW HAVE $   123


        *** CIEL MAGFIL        IS COMING OUT ***


SLICK HANS , NAME YOUR BET ?DON'T PASS

HOW MUCH DO YOU WANT TO BET ?24


ON THE COME-OUT ROLL, CIEL MAGFIL        ROLLS

   3     1

POINT  4

DO YOU WANT TO LAY DOUBLE ODDS AT A PAY-OFF 3     TO 6?YES

NOW COME THE POINT ROLLS
   5     4

   4     2

   5     6

   5     1

   5     4

   3     2

   6     5

   1     2

   6     6
```

```
  3     3

  1     6
```

```
CIEL MAGFIL       SEVENS OUT AND
CHICAGO SAM       IS THE NEXT SHOOTER.
```

YOU WON $  48   AND YOU NOW HAVE $   171


              *** CHICAGO SAM       IS COMING OUT ***

SLICK HANS , NAME YOUR BET ?DON'T PASS

HOW MUCH DO YOU WANT TO BET ?36


ON THE COME-OUT ROLL, CHICAGO SAM       ROLLS
```
  5     3
```
POINT  8

DO YOU WANT TO LAY DOUBLE ODDS AT A PAY-OFF 5     TO 6?YES

NOW COME THE POINT ROLLS
```
  2     3

  5     1

  5     6

  6     6

  6     6

  6     6

  4     5

  6     4

  2     4

  3     4
```
```
CHICAGO SAM       SEVENS OUT AND
LITTLE JIM        IS THE NEXT SHOOTER.
```

YOU WON $  96   AND YOU NOW HAVE $   267


              *** LITTLE JIM       IS COMING OUT ***

SLICK HANS , NAME YOUR BET ?DON'T PASS

HOW MUCH DO YOU WANT TO BET ?90

```
ON THE COME-OUT ROLL, LITTLE JIM            ROLLS

   5      4

POINT  9

NOW COME THE POINT ROLLS
   4      1

   4      5

POINT !

YOU LOST $  90    AND YOU NOW HAVE $  177
```

# Chapter 5

# blackjack

```
************
* BLACKJACK *
************

  * THE DEALER MUST STAND ON 17 AND MUST DRAW TO 16 *

 *** 1 DOLLAR MINIMUM BETS, 500 DOLLAR MAXIMUM BETS ***

   * NO DOUBLING DOWN OR SPLITTING OF SPLIT HANDS *

        *** ONLY ONE DRAW EACH TO SPLIT ACES ***

              * BLACKJACK PAYS 3 TO 2 *

PLEASE IDENTIFY YOURSELF ?HANS SAGAN

HOWDY HANS, YOU WILL PLAY AGAINST JASON THE
CYCLOPS, AND YOU AIN'T SEEN NOTHIN' YET !

HANS, HOW MUCH MONEY ARE YOU WILLING TO PUT UP ?1000

JASON IS GOING TO MATCH THIS AMOUNT.

       *** STAND BY - THE CARDS ARE BEING SHUFFLED ***

  *** PLEASE CUT - BY ENTERING A NUMBER BETWEEN 1 AND 208 ***

?113
THE FIRST CARD, NAMELY THE          *** SEVEN OF CLUBS    ***
IS BEING BURNED.

            *** HIGH CARD GETS TO DEAL ***
```

*Fig. 5.1 Run of the program "Blackjack"*

© 1958 United Features Syndicate, Inc.

```
HERE IS THE

*** FOUR   OF DIAMONDS ***

FOR HANS, AND HERE IS THE           *** JACK   OF SPADES    ***

FOR JASON.


             *** JASON IS THE DEALER ***


HANS, HOW MUCH DO YOU WANT TO BET ?20


HANS,YOUR HOLE CARDS ARE

*** FIVE  OF SPADES    ***
*** SEVEN OF HEARTS    ***

AND JASON'S UP-CARD IS              *** THREE OF CLUBS     ***


DO YOU WANT A HIT ?NO


JASON'S HOLE CARD IS                *** ACE   OF HEARTS    ***
JASON DRAWS THE                     *** FIVE  OF DIAMONDS ***

JASON WINS WITH A COUNT OF  19    VS. HANS'S COUNT OF   12

ONCE THE BETS ARE SETTLED, HERE IS HOW YOU STAND:

                    HANS    $  980      ( -20   )
                    JASON   $ 1020      (  20   )


HANS, HOW MUCH DO YOU WANT TO BET ?40


HANS,YOUR HOLE CARDS ARE

*** EIGHT OF SPADES    ***
*** SIX   OF DIAMONDS ***
```

Fig. 5.1 Run of the program "Blackjack" (cont'd)

```
AND JASON'S UP-CARD IS              *** TEN   OF SPADES   ***


DO YOU WANT A HIT ?YES
HERE IS THE
*** SIX   OF HEARTS   ***
DO YOU WANT A HIT ?NO


JASON'S HOLE CARD IS               *** NINE  OF DIAMONDS ***

JASON LOSES WITH A COUNT OF  19    VS. HANS'S COUNT OF  20

ONCE THE BETS ARE SETTLED, HERE IS HOW YOU STAND:

              HANS      $ 1020     (  40   )
              JASON     $  980     ( -40   )


HANS, HOW MUCH DO YOU WANT TO BET ?80


HANS,YOUR HOLE CARDS ARE

*** EIGHT OF CLUBS     ***
*** EIGHT OF HEARTS    ***

AND JASON'S UP-CARD IS              *** NINE  OF CLUBS    ***


HANS ,DO YOU WANT TO SPLIT YOUR HAND ?YES
HERE IS THE
*** QUEEN OF SPADES    ***
TO YOUR
*** EIGHT OF CLUBS     ***
DO YOU WANT A HIT ?NO



HERE IS THE
*** FOUR  OF HEARTS    ***
TO YOUR
*** EIGHT OF HEARTS    ***
DO YOU WANT A HIT ?YES
HERE IS THE
*** EIGHT OF HEARTS    ***
DO YOU WANT A HIT ?NO



JASON'S HOLE CARD IS               *** KING  OF DIAMONDS ***


JASON LOSES WITH A COUNT OF  19    VS. HANS'S COUNT OF  20
ON THE SECOND HAND AND

JASON WINS WITH A COUNT OF  19     VS. HANS'S COUNT OF  18
ON THE FIRST HAND.

HANS, HOW MUCH DO YOU WANT TO BET ?160
```

Fig. 5.1 Run of the program "Blackjack" (cont'd)

```
HANS,YOUR HOLE CARDS ARE

*** THREE OF DIAMONDS ***
*** FIVE  OF CLUBS    ***

AND JASON'S UP-CARD IS              *** FIVE   OF HEARTS   ***


DO YOU WANT TO DOUBLE DOWN ?YES
HERE IS THE
*** TWO   OF DIAMONDS ***

JASON'S HOLE CARD IS                *** ACE   OF HEARTS   ***
JASON DRAWS THE                     *** QUEEN OF HEARTS   ***
JASON DRAWS THE                     *** JACK  OF HEARTS   ***


JASON IS BUSTED WITH A COUNT OF  26

ONCE THE BETS ARE SETTLED, HERE IS HOW YOU STAND:

                    HANS    $  1340    (  320  )
                    JASON   $   660    ( -320  )


HANS, HOW MUCH DO YOU WANT TO BET ?200


HANS,YOUR HOLE CARDS ARE

*** TEN   OF SPADES   ***
*** ACE   OF SPADES   ***

AND JASON'S UP-CARD IS              *** ACE   OF CLUBS    ***


           ********** BLACKJACK **********

JASON'S HOLE CARD IS                *** NINE  OF DIAMONDS ***


JASON LOSES WITH A COUNT OF  20    VS. HANS'S COUNT OF  21

ONCE THE BETS ARE SETTLED, HERE IS HOW YOU STAND:

                    HANS    $  1640    (  300  )
                    JASON   $   360    ( -300  )


           *** HANS IS THE DEALER ***


JASON BETS  10


HANS, YOUR UP-CARD IS               *** QUEEN OF CLUBS    ***
AND YOUR HOLE CARD IS               *** FOUR  OF SPADES   ***

JASON DRAWS THE
*** FOUR  OF SPADES   ***
AND THE
```

*Fig. 5.1 Run of the program "Blackjack" (cont'd)*

```
*** THREE OF HEARTS    ***
AND THE
*** TEN    OF SPADES    ***
TO THE
*** TWO    OF DIAMONDS ***
*** THREE OF HEARTS    ***
```

JASON IS BUSTED WITH A COUNT OF   22

ONCE THE BETS ARE SETTLED, HERE IS HOW YOU STAND:

```
                HANS    $  1650    (  10   )
                JASON   $   350    ( -10   )
```

JASON BETS   10


HANS, YOUR UP-CARD IS              *** KING   OF DIAMONDS ***
AND YOUR HOLE CARD IS              *** TWO    OF DIAMONDS ***

JASON IS GOING TO SPLIT HIS HAND.
HERE IS THE
*** TEN    OF CLUBS    ***
TO JASON'S
*** ACE    OF HEARTS   ***

HERE IS THE
*** FOUR   OF HEARTS   ***
TO JASON'S
*** ACE    OF SPADES   ***


HANS, DO YOU WANT A HIT ?YES


HERE IS THE                        *** EIGHT OF SPADES    ***


HANS WINS WITH A COUNT OF   20    VS. JASON'S COUNT OF   15
ON THE SECOND HAND AND

HANS LOSES WITH A COUNT OF   20    VS. JASON'S COUNT OF   21
ON THE FIRST HAND.



JASON BETS   10


HANS, YOUR UP-CARD IS              *** FOUR   OF CLUBS    ***
AND YOUR HOLE CARD IS              *** KING   OF SPADES   ***

JASON DRAWS THE
*** TWO    OF HEARTS   ***
TO THE
*** TEN    OF HEARTS   ***
*** TWO    OF CLUBS    ***


HANS, DO YOU WANT A HIT ?YES
```

Fig. 5.1 Run of the program "Blackjack" (cont'd)
```

```
HERE IS THE                          *** JACK  OF SPADES   ***


HANS IS BUSTED WITH A COUNT OF  24

ONCE THE BETS ARE SETTLED, HERE IS HOW YOU STAND:

                    HANS     $   1640     (  -10    )
                    JASON    $    360     (   10    )


JASON BETS   20


HANS, YOUR UP-CARD IS                *** QUEEN OF SPADES   ***
AND YOUR HOLE CARD IS                *** SEVEN OF SPADES   ***

JASON DRAWS THE
*** FOUR   OF HEARTS   ***
TO THE
*** SIX    OF HEARTS   ***
*** ACE    OF HEARTS   ***


HANS LOSES WITH A COUNT OF  17    VS. JASON'S COUNT OF   21

ONCE THE BETS ARE SETTLED, HERE IS HOW YOU STAND:

                    HANS     $   1620     (  -20    )
                    JASON    $    380     (   20    )


JASON BETS   20


HANS, YOUR UP-CARD IS                *** EIGHT OF CLUBS    ***
AND YOUR HOLE CARD IS                *** TWO   OF CLUBS    ***

JASON IS GOING TO SPLIT HIS HAND.
HERE IS THE
*** FOUR   OF HEARTS   ***
TO JASON'S
*** SEVEN OF DIAMONDS ***
JASON DRAWS THE
*** NINE   OF DIAMONDS ***

HERE IS THE
*** NINE   OF HEARTS   ***
TO JASON'S
*** SEVEN OF CLUBS    ***
JASON DRAWS THE
*** SEVEN OF SPADES   ***


HANS, DO YOU WANT A HIT ?YES


HERE IS THE                          *** KING   OF CLUBS   ***



JASON IS BUSTED WITH A COUNT OF  23    ON THE SECOND HAND AND
```

Fig. 5.1 Run of the program "Blackjack" (cont'd)

```
HANS PUSHES WITH A COUNT OF  20    VS. JASON'S COUNT OF  20
ON THE FIRST HAND.
```

```
ONCE THE BETS ARE SETTLED, HERE IS HOW YOU STAND:

                    HANS    $  1640     (  20   )
                    JASON   $   360     ( -20   )
```

```
JASON BETS  10
```

```
HANS, YOUR UP-CARD IS              *** KING   OF HEARTS   ***
AND YOUR HOLE CARD IS              *** ACE    OF CLUBS    ***
```

```
                ********** BLACKJACK **********
```

```
JASON'S HOLE CARDS ARE

*** QUEEN OF HEARTS    ***
*** QUEEN OF CLUBS     ***

HANS WINS WITH A COUNT OF  21    VS. JASON'S COUNT OF  20

ONCE THE BETS ARE SETTLED, HERE IS HOW YOU STAND:

                    HANS    $  1650     (  10   )
                    JASON   $   350     ( -10   )
```

```
JASON BETS  10
```

```
HANS, YOUR UP-CARD IS              *** FOUR   OF HEARTS   ***
AND YOUR HOLE CARD IS              *** TWO    OF SPADES   ***
```

```
                ********** BLACKJACK **********
```

```
JASON'S HOLE CARDS ARE

*** ACE    OF DIAMONDS ***
*** TEN    OF CLUBS    ***

HANS LOSES WITH A COUNT OF   6    VS. JASON'S COUNT OF  21

ONCE THE BETS ARE SETTLED, HERE IS HOW YOU STAND:

                    HANS    $  1635     ( -15   )
                    JASON   $   365     (  15   )
```

```
            *** JASON IS THE DEALER ***
```

Fig. 5.1 Run of the program "Blackjack" (cont'd)

```
HANS, HOW MUCH DO YOU WANT TO BET ?100


HANS,YOUR HOLE CARDS ARE

*** SIX    OF DIAMONDS ***
*** ACE    OF CLUBS    ***

AND JASON'S UP-CARD IS            *** SIX    OF HEARTS   ***


DO YOU WANT TO DOUBLE DOWN ?YES
HERE IS THE
*** ACE    OF HEARTS   ***

JASON'S HOLE CARD IS              *** TEN    OF HEARTS   ***
JASON DRAWS THE                   *** JACK   OF DIAMONDS ***


JASON IS BUSTED WITH A COUNT OF   26

ONCE THE BETS ARE SETTLED, HERE IS HOW YOU STAND:

             HANS    $  1835    (  200  )
             JASON   $  165     ( -200  )


HANS, HOW MUCH DO YOU WANT TO BET ?165


HANS,YOUR HOLE CARDS ARE

*** ACE    OF DIAMONDS ***
*** TWO    OF SPADES   ***

AND JASON'S UP-CARD IS            *** TWO    OF DIAMONDS ***


DO YOU WANT TO DOUBLE DOWN ?NO
DO YOU WANT A HIT ?YES
HERE IS THE
*** FIVE  OF CLUBS    ***
DO YOU WANT A HIT ?NO


JASON'S HOLE CARD IS              *** NINE  OF HEARTS   ***
JASON DRAWS THE                   *** FOUR  OF HEARTS   ***
JASON DRAWS THE                   *** QUEEN OF SPADES   ***


JASON IS BUSTED WITH A COUNT OF   25

ONCE THE BETS ARE SETTLED, HERE IS HOW YOU STAND:

             HANS    $  2000    (  165  )
             JASON   $  0       ( -165  )


CONGRATULATIONS ! YOU WHIPPED JASON REAL GOOD !

DONE
```

Fig. 5.1 Run of the program "Blackjack" (cont'd)

## HOW THE GAME IS PLAYED

Blackjack is probably the most popular and most widely played banking card game in the United States. It is played in gambling houses, private clubs, political club rooms, barracks, troop transports, back rooms of all kinds, and places you may never have heard of. The game is less popular in Europe. Incidentally, the French call it *vingt-et-un* (twenty-one), not *Jacques noir*, and the Bavarians call it *oana-zwoanzgerln*, which, according to expert dialecticians, also means twenty-one.

In view of the game's popularity, it seems almost ludicrous and presumptuous to try to explain it, were it not for the fact that most people play it atrociously, giving the dealer an abnormal advantage that he would not otherwise have. In fact, the player's expected value is positive (Ref. [2], p. 234, 245), and blackjack appears to be the only casino game offering that advantage!

In many respects, blackjack resembles chemin-de-fer (see Chap. 3). It is a card game; the dealer-banker plays an exceptional role; the count of one's hand determines the outcome of the game; and there even are "naturals." In blackjack, however, *every* player is dealt a hand; the count is *not* modulo 10; the ace is *double valued*, counting 1 or 11; and finally, the player may take as many hits as he wants.

Furthermore, the language at the blackjack table is usually less restrained than at the chemin-de-fer or baccarat table (you may hear an occasional "*damn it*" or worse, as opposed to an exasperated and elegantly mouthed "*mon dieu*" at the chemin-de-fer table), and, generally speaking, the stakes tend to be smaller. It appears that while the stakes on the Las Vegas Strip range from a $1 minimum to a $500 maximum, there are also places where one can play for as little as two bits. (In Atlantic City, the table minima are $2, $5, and $25, and the table maximum is usually $500.)

Here is how the game goes. It is played at a table as depicted in Fig. 5.2. The houseman sees to it that one to four decks of ordinary playing cards get ceremoniously shuffled and cut, a ritual in which the players are invited to participate. He then asks for bets, which the players place before them in full view of the dealer, and proceeds to deal a card face down to every one of one to seven players, in a clockwise direction. He himself takes his first card up. (Check for local variations. In some casinos, the player's cards are dealt face up and the dealer helps the player count. How players who cannot even count ever got the money to gamble transcends the author's comprehension.) The dealer proceeds to deal a second card "down and dirty" to everyone including himself. The cards count their face value, court cards count 10, and the ace counts 1 or 11, whichever is of greater advantage to the player. (For more details on the mechanics of the

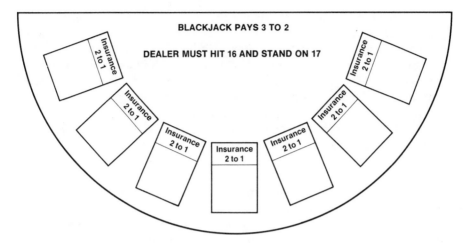

*Fig. 5.2 Blackjack table*

game, see Ref. [10], p. 138ff). From now on, we will use 10 to denote court cards as well as 10's. If the dealer's up card is a 10 or an ace, he looks at his down card immediately. If that is an ace (to a 10) or a 10 (to an ace), giving him a "natural" count of 21, he hollers "blackjack" and rakes in all the stakes from all the players, except the ones who also have blackjack. These lucky ones may retrieve their stake. The round is over and a new deal commences.

If the dealer does not have a blackjack, but one or more of the players does, then those holding a blackjack are paid off 3:2 (that is, a bet of $200 will get them $300), and for them the round is over. (In the olden days, an extra bonus was awarded to the player if he held the ace of spades and a black—club or spade—jack; hence the name of the game.) The others wait until all the excitement has subsided, and then the real game begins.

It is the objective of the game to get a count as close to 21 as possible but not to exceed 21. The player on the dealer's left is asked if he wants a hit for as many times as he may want to draw additional cards to improve his hand. When he is satisfied, he signifies so by saying "I stay" or "pass," or simply by shaking his head. (In a high class game, players say "yes please" or "no, thank you.") If the player's count goes over 21, he is busted. He has to turn his cards up, and the dealer rakes in his stakes. Some players take a bust in stride, hardly batting an eyelid. Others go into a tailspin, disgorging a torrent of expletives. Once the first player is either satisfied or busted, the dealer turns to the next player, and so on, until all players have been taken care of. Then it is the dealer's turn. His moves are prescribed by very strict rules.

First of all, he turns up his hole card. If his count is seventeen or more, he must stay. If his count is sixteen or less, he must draw until his count reaches at least seventeen. He *has to* count an ace as 11 if this brings his total to a count of 17 to 21, and he has to count it as a 1 otherwise. (Suppose his hole cards are an ace and a 5, giving him a count of 6. If he draws a deuce, this will immediately give him a count of 18 and he will have to stay.) In some places, the dealer draws to a count of 17 if his hand includes an ace that counts 11, giving the house a definite advantage (Ref. [5], p. 99). As soon as the dealer has completed his hand, he pays off all the players that have a higher count than he (without being busted) and rakes in the stakes from the ones with a lower count or a bust hand. The pay-off is even money, that is, equal to the amount that has been bet. If a player's count matches the dealer's, it is a stand-off, and the player may retrieve his stake or let it ride for the next round (check for local deviations).

This is essentially all there is to it, except that the player usually has three other options to playing his hand straight for even money. He may "split" a pair, he may "double down," or he may place an "insurance bet."

### Splitting

When the player is dealt a pair—that is, two cards with the same count (2–2, A–A, 8–8, etc.)—he may elect to "split" his hand by placing both cards face up, taking an automatic hit on each one, and playing each group of two cards as separate hands by putting a stake equal to the original one on the second hand. In fact, the player acts now as two players. He may win with one hand and get busted with the other; lose both; win one and lose one; get busted with both; get busted with one and lose one honestly; or win both. If one of his cards gets hit by another card of the same value, he may split that hand again and wind up playing three hands so long as he triples his original bet. In principle, this could go on and on; in practice, it does not. Note that when you split aces, you are allowed only one hit for each hand. If it is a court card or a 10, you do *not* have a blackjack, just an ordinary count of 21! Also note that 10–jack, queen–king, etc., are also considered pairs, but it is *not* in the player's interest to split 10's because a count of 20 is the best he is likely to get.

### Doubling Down

The player may elect to double his bet once he has seen his two-card hand. In exchange for this privilege, he must agree to turn his two cards up and accept only one hit, which will be face down. This process is called *doubling down*. A player may double down on all split hands if he chooses.

## Insurance Bet

When the dealer's up card is an ace and you don't have a blackjack, you may "insure" your hand against the possibility of losing to the dealer's potential blackjack. The dealer offers such a bet *before* looking at his down card. It costs you half your original bet and pays off 2:1, that is, if the dealer does have a blackjack, then, regardless of what you have, you win an amount equal to your original wager while losing your original stake and thus break even. An insurance bet is a sucker bet and is not recommended (unless you somehow know that more than one-third of the unseen cards have value 10). Normally, the odds against winning an insurance bet are higher than 2:1 (Ref. [14], p. 57). Some casinos offer an insurance bet at 10:1 against the dealer's down card being an ace if his up card is a 10 (Ref. [2], p. 237).

## The Player's Strategy

What will be said in this section applies to a game in which the rigid requirements explained in the previous section are placed on the dealer and in which a sufficient number of decks are used to make "counting" difficult and to give, for all practical purposes, every card an equal chance to come up. Edward Thorp (Ref. [13]) was apparently the first player to recognize that the odds change as the cards are played out from a deck and that one can use this fact to one's advantage. He proceeded to develop methods for recalculating the odds before every bet and of playing one's hand accordingly. The method is called *counting*, and his disciples are known as *counters*. It did not take casinos long to catch onto the practice—the counters kept winning—and they put countermoves in motion: Blackjack is now played with more than one deck (some places use as many as four decks) and houseman are instructed to speed up the deal and accompany it with an incessant chatter to break the counter's concentration. (Counters are easily spotted: They have a sort of inward look about them and move their lips ever so slightly!) More about counting later on in this chapter.

When the game is played with four decks, which are reshuffled when 30 or so cards are left, one may assume that each card has about the same chance of being drawn at any given time, this chance being 1 out of 13 if each court card is counted separately. (The chance of a card with a count of 10 coming up is, of course, 4 out of 13.)

To explain strategy, it is advantageous to avail oneself of the pertinent technical language. The following terms should be part of every blackjack player's vocabulary:

*Soft hand*    A soft hand is a hand in which an ace is counted as 11 so long as the total count does not exceed 21 (for example, A–3–6 = 20).

*Hard hand*     A hard hand is one without aces or a hand with an ace (or aces) which has (have) to be counted as a 1 because the count would otherwise exceed 21 (for example, 3–Q–5 = 18; J–8–A = 19).

Note that only one ace can be counted as 11 because two aces counted as 11 would yield a count of 22. In principle, the lowest count of a soft hand would be 12 (for two aces), but since one *always* splits two aces, the case never arises.

*Stiff hand*     A stiff hand is a hard hand with a count of 12 to 16. If you draw to a stiff hand, you run the risk of getting busted. (For example, a 10 to a count of 12 will do it, or a 6 to a count of 16.) Of course, you can also get busted when drawing to a count of 17 or more, but you don't ever consider drawing to a hard 17—unless you are a masochist.

*Good cards and poor cards*     A good card is a 7, 8, 9, 10, or ace, and a *poor card* is a 2, 3, 4, 5, or 6.

In a friendly, relaxed game (where the principal objectives are drinking and telling bawdy stories) for small stakes (such as dimes), the following simple strategy may suffice:

1. Stay with a hard hand of 17 or more.
2. Stay with a soft hand of 19 or more.
3. Draw to a stiff hand if the dealer shows a good card.
4. Stay with a stiff hand if the dealer shows a poor card.
5. Draw to a soft hand of 18 or less.
6. Draw to a hard hand of 11 or less.

However, if you are serious about the game and want to make some money while having fun, you had better follow—at the very least—the more complicated *optimal (zero-memory) strategy* that has been developed by Epstein (Ref. [2], p. 245ff). Epstein arrived at his strategy by calculating the player's expectations for each of the 550 distinct initial situations. (There are 55 distinct two-card hands, and the dealer may show one of ten possible cards, making 550 different situations altogether.)

You examine your hand, look at the dealer's up card, and then make your decision, as follows:

If you are dealt a pair, you seriously consider splitting your hand. You consult Table 5.1 and follow instructions. (Note that one never splits 5's and 10's. A pair of 5's is a good hand to double down—unless the dealer shows a 10 or an ace—and a count of 20 cannot easily be improved.)

Initial situations involving a pair and not calling for a split will have to be checked in Table 5.2 for exceptional hands (there are eight

**Table 5.1  Splitting of pairs**

| PLAYER'S HAND | DEALER'S UP CARD | | | | | | | | | |
|---|---|---|---|---|---|---|---|---|---|---|
| | 2 | 3 | 4 | 5 | 6 | 7 | 8 | 9 | 10 | A |
| 2-2 | ▨ | ▨ | ▨ | ▨ | ▨ | ▨ | | | | |
| 3-3 | ▨ | ▨ | ▨ | ▨ | ▨ | ▨ | | | | |
| 4-4 | | | | ▨ | | | | | | |
| 5-5 | | | | | | | | | | |
| 6-6 | ▨ | ▨ | ▨ | ▨ | ▨ | | | | | |
| 7-7 | ▨ | ▨ | ▨ | ▨ | ▨ | ▨ | ▨ | | * | |
| 8-8 | | ▨ | ▨ | ▨ | ▨ | ▨ | ▨ | ▨ | ▨ | ▨ |
| 9-9 | ▨ | ▨ | ▨ | ▨ | ▨ | | ▨ | ▨ | | |
| 10-10 | | | | | | | | | | |
| A-A | ▨ | ▨ | ▨ | ▨ | ▨ | ▨ | ▨ | ▨ | ▨ | ▨ |

* CHECK FOR POSSIBLE EXCEPTION  ▨ SPLIT  ☐ CHECK UNDER HARD HAND

**Table 5.2  Exceptional hands**

| PLAYER'S HAND | DEALER'S UP CARD | | | | | | | | | |
|---|---|---|---|---|---|---|---|---|---|---|
| | 2 | 3 | 4 | 5 | 6 | 7 | 8 | 9 | 10 | A |
| 2-6 | | | | | ST | | | | | |
| 2-10 | | | D | | | | | | | |
| 3-9 | | ST | | | | | | | | |
| 4-8 | | ST | | | | | | | | |
| 5-7 | | ST | | | | | | | | |
| 3-5 | | | | DD | | | | | | |
| 3-10 | D | | | | | | | | | |
| 7-7 | | | | | | | | | ST | |

ST Stay   D Draw   DD Double Down

**Table 5.3   Hard hands**

| PLAYER'S TOTAL ＼ DEALER'S UP CARD | 2 | 3 | 4 | 5 | 6 | 7 | 8 | 9 | 10 | A |
|---|---|---|---|---|---|---|---|---|---|---|
| 17 or more | | | | | | | | | | |
| 16 | | | | | | ▨ | ▨ | ▨ | ▨ | ▨ |
| 15 | | | | | | ▨ | ▨ | ▨ | ▨ | ▨ |
| 14 | | | | | | ▨ | ▨ | ▨ | ▨* | ▨ |
| 13 | * | | | | | ▨ | ▨ | ▨ | ▨ | ▨ |
| 12 | ▨ | ▨* | * | | | ▨ | ▨ | ▨ | ▨ | ▨ |
| 11 | ■ | ■ | ■ | ■ | ■ | ■ | ■ | ■ | ■ | ■ |
| 10 | ■ | ■ | ■ | ■ | ■ | ■ | ■ | ■ | ▨ | ▨ |
| 9 | ▨ | ■ | ■ | ■ | ■ | ▨ | ▨ | ▨ | ▨ | ▨ |
| 8 | ▨ | ▨ | ▨ | ▨ | ▨* | ■* | ▨ | ▨ | ▨ | ▨ |
| 7 or less | ▨ | ▨ | ▨ | ▨ | ▨ | ▨ | ▨ | ▨ | ▨ | ▨ |

Legend: **\*** CHECK FOR POSSIBLE EXCEPTION    (blank) STAY    ▨ DRAW    ■ DOUBLE DOWN

exceptional hands that do not fit the general pattern) and in Table 5.3 for hard hands. (A pair that has not been split always constitutes a hard hand, because it has no aces in it. Aces are always split.)

If you are holding a hard hand other than a pair, consult Table 5.3. Some hard hands are exceptional and have to be looked up in Table 5.2. Soft hands are taken care of in Table 5.4. That's it as far as the first draw is concerned.

From now on, things get sticky. There are hundreds of thousands of multiple card situations, and the computation of various strategies becomes counterproductive. Epstein (Ref. [2], p. 249ff) has carried out the computations for a number of cases and arrived at some rules such as the following: If the dealer shows a 10, draw to a three- or four-card hand with a 16 count if you have two 6's; a 6 with a 7, 8, or 9; or three deuces and a 10. If the dealer shows a 7, stay with a four-card hand of 16 if your highest card is a 5 or less (ace counts 1). Our advice to beginners: For the second draw, follow the simple strategy listed previously, or, if you must, follow the same rules as before as far as they are still applicable. (Note that you cannot double down on three cards or more; you'll have to take a hit instead. On the other hand, if

**Table 5.4  Soft hands**

| DEALER'S UP CARD / PLAYER'S TOTAL | 2 | 3 | 4 | 5 | 6 | 7 | 8 | 9 | 10 | A |
|---|---|---|---|---|---|---|---|---|---|---|
| 19 or more | | | | | | | | | | |
| 18 | | | ■ | ■ | ■ | | | ▨ | ▨ | ▨ |
| 17 | | ■ | ■ | ■ | ■ | ▨ | ▨ | ▨ | ▨ | ▨ |
| 16 | ▨ | ▨ | ■ | ■ | ■ | ▨ | ▨ | ▨ | ▨ | ▨ |
| 15 | ▨ | ▨ | ■ | ■ | ■ | ▨ | ▨ | ▨ | ▨ | ▨ |
| 14 | ▨ | ▨ | ■ | ■ | ■ | ▨ | ▨ | ▨ | ▨ | ▨ |
| 13 | ▨ | ▨ | ■ | ■ | ■ | ▨ | ▨ | ▨ | ▨ | ▨ |

☐ STAY    ▨ DRAW    ■ DOUBLE DOWN

you split your hand initially, the same rules apply if you are dealt a third card with the same count!)

Some casinos (at least one in Atlantic City) offer the so-called *surrender option*. Under this option, you may relinquish a "hopeless" hand if the dealer's up card is not an ace, forfeiting half your wager. According to Epstein (Ref. [2], p. 250), the following two-card hands should be surrendered if the dealer's up card is a 10: 5–10, 6–10, 6–9, 7–9, and 7–7.

Just as you can develop an optimal strategy for playing with a full deck (or full decks) of cards, you can also develop one for the case where certain cards are missing. Thorp's method of *counting fives* (Ref. [13], p. 47ff) is based on switching to an optimal strategy for playing blackjack without 5's as soon as all the 5's have been dealt. To do so will tilt the game in your favor to the extent of expected winnings of $30 per $1000 bet. Thorp's *ten-count strategy* (Ref. [13], p. 102ff) consists essentially in varying your bets according to the ratio of 10's versus non-10's that are still in the deck and gives the player an advantage of expected winnings from $10 to $100 per $1000 bet! For those who intend to play a great deal of blackjack without unlimited resources, it may be worthwhile to study Thorp's strategies. His systems are based on sound mathematical principles, as opposed to the fanciful systems discussed in Chap. 2 in conjunction with roulette, which are based on mere wishful thinking!

## THE COMPUTER PROGRAM

When blackjack is played privately among friends rather than in a gaming establishment, the deal is passed along among the players.

Who gets to deal first is usually decided by a dealer "pro tem" who deals cards face up in the clockwise direction. Whoever gets the first ace gets to deal. As an alternative, it may be decided that high card gets to deal. (If there is a draw, the deal is repeated until there is a decision.) The dealer keeps the deal until one player has a blackjack (and the dealer does not). If more than one player is that lucky, then the one nearest to the dealer's left gets the deal. (A blackjack, incidentally, appears about twice every 43 hands or so.) In a friendly game, the pay-off on a blackjack is customarily 2:1. What to do in case a player refuses to accept the deal is left for the group to decide. After all, this is supposed to be a friendly game! To provide for a more stable situation, Scarne (Ref. [10], p. 140) suggests that the deal should be passed on after the completion of five deals regardless of the occurrence of blackjacks.

When writing the computer program for *blackjack* that produced the printout in Fig. 5.1, we decided to make it a friendly game that may easily be converted into a casino game. We let the operator play against the computer with the deal passing between them according to the rules outlined in the preceding paragraph. We did, however, retain the casino's 3:2 pay-off on a blackjack. High card will get to deal first. There is no point in including fictional players (as we did in Chap. 3 for chemin-de-fer) because every one of them would have to be dealt a two-card hand face down, and there is not much point in watching the computer play against itself while revealing very little of what is actually going on. When $T = 1$ (see line 620, Fig. 5.3), the computer is the dealer; when $T = 2$ (see line 680), the operator is the dealer.

We play the game with four decks of cards and reshuffle when we are down to about 30 cards. We do not permit the splitting of split hands (which won't occur all that often anyway—the author has seen the circumstances that would call for the splitting of a split hand only once. The dolt was dealt two deuces to the dealer's 7. He didn't split his hand, drew a third deuce, then an 8, stayed, and lost all of his two dollars!). We also won't permit doubling down on split hands in order to keep a complicated program from getting even more complicated.

The casino version, where the computer acts as dealer, may easily be obtained from our program with two GOTO statements that we shall discuss later. To accommodate the splitting and doubling down of split hands is a little more complicated but poses no real problem for the experienced programmer.

When the computer acts as dealer, it follows the strict rules previously laid down for completing its hand. As the player, the computer is programmed to follow Epstein's optimal (zero-memory) strategy for the first draw and the simple strategy on p. 160 for all following draws.

As in all our programs, the computer records bets and takes care of all the bookkeeping chores. Whenever the computer is the player, it places its bets at random, but within some very reasonable limits. We do not allow any bets that exceed the player's capability to pay or the dealer's ability to back but we do permit "overruns" in case the player loses on doubling down or splitting hands when the bets are, in effect, doubled. It can also happen that the dealer may go into the red when paying off on a blackjack. To avoid this situation would have further complicated an already very complicated program and would not really have served any useful purpose.

To begin with, we have to shuffle and cut four decks of cards. We use the same procedure (lines 4080 to 4420 in Fig. 5.3) that we used in Chaps. 1 and 3, modified to accommodate four rather than six decks. (See also our remark on p. 15.)

Next we have to keep track of how many cards have been played or burned. We accomplish this task by the following device: We let P denote the number of cards that have been used at the time a new deal begins. At this point, the player is dealt cards number P+1 and P+3. The dealer's up card is card number P+2, and his hole card is card number P+4. We use N to count the number of cards that have been displayed since the inception of the new deal, advancing N by 1 every time the name of a card is printed out. Since this print-out has to be accessible from several points in our program, we put it in a subroutine:

```
3620   PRINT "*** "V$[1+5*(V[J]-1),5+5*(V[J]-1)]" OF ";
3630   PRINT S$[1+8*L[J],8+8*L[J]]" ***"
3640   N=N+1
3650   RETURN
```

[V\$, S\$, V(J), and L(J) have the same meaning as they had in Chaps. 1 and 3.] It can happen that the dealer's down card never gets displayed and hence won't get counted: When the operator is the player and his hand is busted in case of a straight play, or both his hands are busted in case of a split hand, the dealer's hole card is burned and slips by the count. We compensate for this by advancing N elsewhere in the program (see lines 1780 and 2130, remembering that T = 1 means that the operator is the player and T = 2 means that the operator is the dealer). At the time that the next deal commences, N+P cards have been played since the beginning of the game, or since the last reshuffle (see line 950).

The count of the player's and the dealer's hands is of the utmost importance for every aspect of the game. This task is rendered somewhat more difficult by the fact that not only is the ace double-valued but is counted in a different manner by dealer and player. (Remember that the dealer counts an ace as 11 if that brings his total to 17 or more, but not more than 21, but the player counts an ace as 11 as long as it keeps his total from exceeding 21. In the final analysis, it does not really matter, but that's the way they count, and we try to give as realistic a simulation as possible, never mind whether it makes sense or not!)

We use the variable G0 for a player's hard count and G1 for a player's soft count, with the understanding that G0 = G1 if the hand has no aces or if the counting of an ace as 11 would cause the total count to exceed 21 (G stands for "gambler"). Similarly, we use H0 to denote the dealer's hard count and H1 to denote the dealer's soft count (H stands for "house"). We let H0 = H1 if there are no aces in the dealer's hand or if the counting of an ace as 11 would either cause the total to exceed 21 or would not bring the count to 17, 18, 19, 20, or 21.

In either case, we work with only G1 and H1 once all the cards have been drawn by player as well as dealer. (If G1, H1 are less than or equal to 21, then they represent the more advantageous count anyway, and they are only over 21 if G0, H0 are also over 21. Remember that as soon as G1 or H1 exceeds 21, it is set back to G0 or H0, respectively.)

For example, if a player holds 2–A–7, then G0 = 10 and G1 = 20. If the dealer holds 4–A–A, then H0 = 6 and H1 = 6, but if he holds 5–A–A, then H0 = 7 and H1 = 17.

Since an evaluation of the player's hand as well as the dealer's hand has to be accessible from many points of the program, we put these evaluations into subroutines. We evaluate the player's first two cards by means of

```
3700   G0=G[P+1]+G[P+3]
3710   IF G[P+1] <> 1 AND G[P+3] <> 1 THEN 3740
3720   G1=G0+10
3730   RETURN
3740   G1=G0
3750   RETURN
```

and the dealer's first two cards with

```
3840   H0=G[P+2]+G[P+4]
3850   IF G[P+2] <> 1 AND G[P+4] <> 1 THEN 3880
3860   IF H0<7 OR H0>11 THEN 3880
3870   GOTO 3900
3880   H1=H0
3890   GOTO 3910
3900   H1=H0+10
3910   RETURN
```

(Note line 3860, which accommodates the dealer's whimsical way of counting his hand.)

From now on, whenever a card is to be drawn, it is either card number J = P+N+1 if the operator is the dealer (in which case both the dealer's cards have been displayed, and P+N gives a true count of the cards that have been used) or number J = P+N+2 if the operator is the player and the dealer's down card is still hidden but already in the pipeline. (In this case, N is one short because it has not yet registered the dealer's down card, although that card has already been "dealt.")

Here is how we count the player's hand after each draw:

```
3760    IF G0=G1 AND G[J]=1 AND G1+G[J]+10 <= 21 THEN 3790
3770    G1=G1+G[J]
3780    GOTO 3800
3790    G1=G1+G[J]+10
3800    G0=G0+G[J]
3810    IF G1 <= 21 THEN 3830
3820    G1=G0
3830    RETURN
```

and here is how we count the dealer's hand after each draw:

```
3920    H0=H0+G[J]
3930    IF G[P+2]=1 OR G[P+4]=1 THEN 3980
3940    FOR I=M TO J
3950    IF G[I]=1 THEN 3980
3960    NEXT I
3970    GOTO 3990
3980    IF H0 >= 7 AND H0 <= 11 THEN 4010
3990    H1=H0
4000    RETURN
4010    H1=H0+10
4020    RETURN
```

(Card number M is the first card the dealer gets to draw after the player has completed his hand. M is set in lines 1670 and 1720 when the operator is the player and in line 3270 when the operator is the dealer. See the displayed program in Fig. 5.3.)

Note the difference in the count of the dealer's hand and the player's hand. If H0 = H1, it does not necessarily mean that none of the cards number P+2, P+4, M, M+1, . . . , J are aces, or, if any one of them is an ace, that the soft count H1 would exceed 21. That is why we have to examine every one of these cards from scratch every time the dealer draws a card. By contrast, G0 = G1 means that none of the cards the player holds is an ace, or, if at least one of them is, then the soft count would exceed 21.

Before going any deeper into an explanation, let us identify the other variables that are bouncing around our program in Fig. 5.3.

S, the "split variable" introduced in line 920, is zero unless the player splits his hand. In the latter case, S = 1 identifies the player's first hand, and S = 2 identifies the player's second hand. H, normally zero, is set to 1 as soon as the result of the play of the second hand has been announced, and, subsequently, because H = 1, the result of the play of the first hand is announced (see lines 930, 2300, and 2340.)

Normally, D, the "double down variable," is zero, unless the player doubles down. Then, D = 1. This will cause the bet to double and also cause exactly one additional card to be dealt to the player's two down cards.

C(1) represents the amount the player wins (negative winnings are losses), except on a split hand. When he does split his hand, then C(1) = 0, and C(2) represents his winnings on the first hand and C(3) represents his winnings on the second hand.

Observe that if the player plays his hand straight and bets B(T), then C(1) = B(T) if he wins and C(1) = −B(T) if he loses. (We remind the reader that T = 1 means that the computer—Jason—is the dealer and T = 2 means that the operator is the dealer. See lines 600–620 and 660–680.) If the player doubles down, then C(1) = ±2*B(T), and if he splits his hand, then C(1) = 0, C(2) = ±B(T), and C(3) = ±B(T). In a draw, C(S + 1) = 0 for whatever S. At the beginning of each deal, C(1), C(2), and C(3) are set to zero in line 1010. If your computer does not have any MAT capabilities, simply replace this line by three assignment statements, one for each C(S+1), where S = 0, 1, 2.

Finally, if the player has a blackjack, then C(1) = 3*B(T)/2. We take care of this in line 2010 by using the variable W, which was introduced in line 1020 and re-evaluated in line 4050. (Clearly, W = 1 if the dealer has a blackjack or nobody does, and W = 2 if the player has a blackjack. Clever, ain't it?)

When the operator is the player and the computer is the dealer, the program does not pose any particular challenge because the dealer's (computer's) actions are rigidly prescribed. (See, for example, Ref. [11], p. 86 for a very "lean" program for blackjack with the computer as dealer.) The computer simply keeps drawing cards until its count reaches 17 and then stops. (See lines 1890ff in Fig. 5.3.)

The operator, acting as player, is a free agent. He is asked in line 1280 if he wants to split his hand, having just been dealt a pair (even if splitting may not be in his best interest). In line 1540, he is asked if he wants to double down, having been dealt a hand with a hard count between 9 and 11, a soft count between 13 and 18, or a trey and a fever versus the computer's fever (see lines 1500–1520). He may or may not wish to do so, for it may not be in his best interest to double down. (We refer the reader to Tables 5.1 through 5.4.) If he elects to split his hand, then the situation gets sticky and is dealt with in lines ·1310 to 1480. If he doubles down, he gets exactly one hit in line 1630. In all other cases, he is asked if he wants a hit provided his count does not exceed 21. If it does exceed 21, the dealer's down card is burned, and the computer announces the bust and takes care of the necessary bookkeeping chores. If the player ceases to draw before getting busted, the computer (dealer) completes its hand (it may get busted in the process), and the two hands are compared. The result is announced in lines 2000 to 2370, and again the bookkeeping chores are attended to.

If the player or the dealer has a blackjack, the subroutine 4030–4070 takes care of the concomitant ballyhoo.

The more interesting part of the program (not necessarily the game) takes place when the computer is the player and the operator is the dealer. For that event, we have incorporated Epstein's optimal (zero-memory) strategy into our program for drawing to the first two cards. The portions of our program that pertain to that strategy are best studied in conjunction with Tables 5.1 through 5.4.

We found it most advantageous to get the pairs out of the way first. To do so, we use the following:

```
2540   IF GCP+1] <> GCP+3] THEN 2650
2550   IF GCP+1]=1 THEN 2630
2560   IF GCP+1]=10 OR GCP+1]=5 THEN 2660
2570   IF GCP+1]=4 AND GCP+2] <> 5 THEN 2660
2580   IF GCP+1] <= 6 AND (GCP+2] >= 8 OR GCP+2]=1) THEN 2660
2590   IF GCP+1]=7 AND (GCP+2] >= 9 OR GCP+2]=1) THEN 2660
2600   IF GCP+1] <> 9 THEN 2630
2610   IF GCP+2] <> 7 AND GCP+2] <> 10 AND GCP+2] <> 1 THEN 2630
2620   GOTO 2660
2630   PRINT P$" IS GOING TO SPLIT HIS HAND."
2640   GOTO 3400
```

(See Table 5.1.) Note that lines 2650 and 2660 initiate the part of the program that checks for exceptional hands and that 3400 initiates the splitting routine.

The exceptional hands (Table 5.2) are dealt with as follows:

```
2650   IF G1 <> G0 THEN 2820
2660   IF G1=13 AND GCP+2]=2 AND (GCP+1]=3 OR GCP+3]=3) THEN 2910
2670   IF G1=12 AND GCP+2]=4 AND (GCP+1]=2 OR GCP+3]=2) THEN 2910
2680   IF G1 <> 12 OR GCP+2] <> 3 THEN 2710
2690   IF GCP+1] >= 3 AND GCP+1] <= 5 THEN 3160
2700   IF GCP+3] >= 3 AND GCP+3] <= 5 THEN 3160
2710   IF G1=8 AND GCP+2]=6 AND (GCP+1]=2 OR GCP+3]=2) THEN 3160
2720   IF G1=8 AND GCP+2]=5 AND (GCP+1]=3 OR GCP+3]=3) THEN 3070
2730   IF G1=14 AND GCP+2]=10 AND GCP+1]=7 THEN 3160
```

Note that the restriction to 63 characters per line that was imposed by the format of this book did cramp our style somewhat. Lines 2680, 2690, and 2700 for example, could have been replaced by one single conjunction. If the count is 12 and the dealer's up-card is a trey and if one of the player's cards is a trey, four, or a fever, then the player should stay. [We had to use deMorgan's law elsewhere as well to break up long conjunctions, by decomposing their negation into a disjunction of negations: NOT ($P_1$ AND $P_2$ AND $P_3$ ... AND $P_n$) is equivalent to NOT $P_1$ OR NOT $P_2$ OR . . . OR NOT $P_n$.] Note that line 2910 leads to a hit, line 3160 to a stay, and line 3070 to doubling down. Since all exceptional hands are hard hands, soft hands are not put through this wringer but passed along from line 2650 to line 2820.

Next, we check the hard hands that have not yet been taken care of either by splitting or as exceptional hands. The following instructions represent the translation of Table 5.3 into programming language:

```
2740   IF G1 >= 17 THEN 3160
2750   IF G1>12 AND GKP+21>1 AND GCP+21<7 THEN 3160
2760   IF G1=12 AND GCP+21>3 AND GCP+21<7 THEN 3160
2770   IF G1=11 THEN 3070
2780   IF G1=10 AND GCP+21 <> 10 AND GCP+21 <> 1 THEN 3070
2790   IF G1=9 AND GCP+21>1 AND GCP+21<7 THEN 3070
2800   IF G1=8 AND GCP+21=6 THEN 3070
2810   GOTO 2910
```

Finally, in

```
2820   IF G1>18 THEN 3160
2830   IF G1 <> 18 THEN 2870
2840   IF GCP+21>2 AND GCP+21<7 THEN 3070
2850   IF GCP+21>8 THEN 2910
2860   GOTO 3160
2870   IF G1=17 AND GCP+21>1 AND GCP+21<7 THEN 3070
2880   IF G1=17 THEN 2910
2890   IF GCP+21<4 OR GCP+21>6 THEN 2910
2900   GOTO 3070
```

we take care of the soft hands to let the computer make its decision according to Table 5.4.

Additional draws are guided by the simple strategy outlined on p. 160. The pertinent instructions are to be found in lines 2950 to 3060. (Watch for the detour for a split hand in line 3000.)

We could have reduced the length of the program somewhat by making more extensive use of the computed GOTO statement, utilizing even more parts of the program jointly for T = 1 and T = 2. We found this approach counterproductive. It only compounds the complexity of our program and tends to confound the reader.

## MODIFICATIONS OF THE COMPUTER PROGRAM

Although it is rare to find oneself in a situation calling for splitting a split hand, it could happen, and the reader may wish to modify the program in order to accommodate such a contingency. Let S = 3,4 characterize the new hands that have been obtained from splitting the first of the two hands (which is characterized by S = 1 and which, in turn, has been obtained from splitting the original hand), and let S = 5,6 characterize the new hands that were obtained from splitting the second of the two hands (which is characterized by S = 2). You'll have to make provisions for a C(3), C(4), C(5), and C(6)

```
10    PRINT TAB(25)"**************"
20    PRINT TAB(25)"* BLACKJACK *"
30    PRINT TAB(25)"**************"
40    DIM S[208],T[208],V[208],L[208],G[208],V$[65],S$[32]
50    DIM A$[3],B$[6],N$[20],D$[12],F$[12],B[2],M[2],C[3]
60    V$[1,30]="ACE   TWO   THREEFOUR FIVE SIX   "
70    V$[31,65]="SEVENEIGHTNINE TEN   JACK QUEENKING "
80    S$="CLUBS    DIAMONDSHEARTS  SPADES   "
90    PRINT LIN(1)
100   PRINT TAB(5)"* THE DEALER MUST STAND ON 17 AND MUST ";
110   PRINT "DRAW TO 16 *"
120   PRINT
130   PRINT TAB(4)"*** 1 DOLLAR MINIMUM BETS, 500 DOLLAR ";
140   PRINT "MAXIMUM BETS ***"
150   PRINT
160   PRINT TAB(7)"* NO DOUBLING DOWN OR SPLITTING OF SPLIT ";
170   PRINT "HANDS *"
180   PRINT
190   PRINT TAB(11)"*** ONLY ONE DRAW EACH TO SPLIT ACES ***"
200   PRINT
210   PRINT TAB(19)"* BLACKJACK PAYS 3 TO 2 *"
220   PRINT LIN(1)
230   PRINT "PLEASE IDENTIFY YOURSELF ";
240   INPUT N$
245   REM THE LOOP 250 TO 270 LOOKS FOR THE BLANK SPACE BETWEEN
246   REM THE OPERATOR'S FIRST NAME AND SURNAME
250   FOR L=0 TO 19
260   IF N$[L+1,L+1]=" " THEN 280
270   NEXT L
280   N$=N$[1,L]
290   PRINT LIN(1)
300   PRINT "HOWDY "N$", YOU WILL PLAY AGAINST JASON THE"
310   PRINT "CYCLOPS, AND YOU AIN'T SEEN NOTHIN' YET !"
320   PRINT LIN(1)
330   PRINT N$", HOW MUCH MONEY ARE YOU WILLING TO PUT UP ";
335   REM M(1), M(2) ARE THE AMOUNTS AT THE OPERATOR'S AND THE
336   REM COMPUTER'S DISPOSAL
340   INPUT M[1]
350   PRINT
360   PRINT "JASON IS GOING TO MATCH THIS AMOUNT."
370   M[2]=M[1]
380   PRINT LIN(1)
390   GOSUB 4080
395   REM N KEEPS COUNT OF THE DISPLAYED CARDS
400   N=0
410   PRINT "THE FIRST CARD, NAMELY THE ";TAB(36);
420   J=1
430   GOSUB 3620
440   PRINT "IS BEING BURNED."
450   PRINT LIN(1)
460   PRINT TAB(16)"*** HIGH CARD GETS TO DEAL ***"
470   PRINT LIN(1)
480   PRINT "HERE IS THE"
490   PRINT
500   J=N+1
510   GOSUB 3620
520   PRINT
530   PRINT "FOR "N$", AND HERE IS THE ";TAB(36);
540   J=N+1
550   GOSUB 3620
560   PRINT
570   PRINT "FOR JASON."
```

*Fig. 5.3 Master program "Blackjack"*

```
580    PRINT LIN(1)
590    GOTO SGN(V[J-1]-V[J])+2 OF 600,640,660
595    REM D$ IS THE DEALER, P$ THE PLAYER. IF T=1, THE COMPUTER
596    REM IS DEALER, IF T=2, THE OPERATOR IS DEALER
600    D$="JASON"
610    P$=N$
620    T=1
630    GOTO 690
640    PRINT "LET'S TRY AGAIN."
650    GOTO 480
660    D$=N$
670    P$="JASON"
680    T=2
690    PRINT LIN(1)
695    REM THE ARGUMENT OF THE TAB FUNCTION IN LINE 700 CENTERS
696    REM THE MESSAGE THAT IS TO BE PRINTED
700    PRINT TAB(17-(LEN(D$)-3)/2);"*** "D$" IS THE DEALER ***"
710    PRINT LIN(1)
720    GOTO 880
730    PRINT LIN(1)
735    REM B(1), B(2) ARE THE PLAYER'S AND THE COMPUTER'S BETS.
736    REM C(1) REPRESENTS THE GAIN OR LOSS FROM PLAYING A
737    REM STRAIGHT HAND, C(2),C(3) FROM PLAYING A SPLIT HAND
740    B[T]=C[1]+C[2]+C[3]
750    IF B[T]=0 THEN 880
755    REM IN LINE 760, THE OPERATOR'S WINNINGS (LOSSES) ARE
756    REM CONVERTED INTO THE COMPUTER'S LOSSES (WINNINGS) AND
757    REM VICE VERSA
760    B[T+1-2*INT(T/2)]=-B[T]
770    M[1]=M[1]+B[1]
780    M[2]=M[2]+B[2]
790    PRINT "ONCE THE BETS ARE SETTLED, HERE IS HOW YOU STAND:"
800    PRINT
810    PRINT TAB(17);N$;TAB(26)"$ "M[1];TAB(38)"( "B[1]")"
820    PRINT TAB(17);"JASON"TAB(26)"$ "M[2];TAB(38)"( "B[2]")"
825    REM IF EITHER PLAYER IS OUT OF MONEY, CONTROL IS PASSED
826    REM FROM LINE 830 TO LINE 4430 TO TERMINATE THE GAME
830    IF M[1] <= 0 OR M[2] <= 0 THEN 4430
840    IF G[P+1]+G[P+3] <> 11 THEN 880
850    IF G[P+1] <> 1 AND G[P+3] <> 1 THEN 880
855    REM IN LINE 860, THE DEAL IS PASSED TO THE PLAYER IF
856    REM HE HAD A BLACKJACK
860    GOTO T*SGN(ABS(H1-G1))+1 OF 880,660,600
870    PRINT LIN(1)
880    IF P<178 THEN 920
890    PRINT TAB(6)"*** STAND BY - ";
900    PRINT "THE CARDS ARE BEING RE-SHUFFLED ***"
910    GOSUB 4100
913    REM FOR A STRAIGHT HAND, S=0. FOR A SPLIT HAND, S=1 FOR
914    REM THE FIRST HAND, S=2 FOR THE SECOND. D, NORMALLY 0,
915    REM IS SET TO 1 IF THE PLAYER DOUBLES DOWN. G0,G1,H0,H1
916    REM ARE THE HARD AND SOFT COUNTS OF THE PLAYER'S HAND
917    REM AND THE DEALER'S HAND, RESPECTIVELY. P IS THE NUMBER
918    REM OF CARDS USED UP AT THE BEGINNING OF THE DEAL, AND
919    REM N KEEPS COUNT OF THE CARDS THAT ARE USED IN THE DEAL
920    S=0
930    H=0
940    D=0
950    G0=0
960    G1=0
970    H0=0
```

Fig. 5.3 Master program "Blackjack" (cont'd)

```
980    H1=0
990    P=N+P
1000   N=0
1010   MAT C=ZER[3]
1015   REM W IS RAISED FROM 1 TO 2 IF THE PLAYER HAS A BLACKJACK
1020   W=1
1030   PRINT LIN(1)
1040   IF T=2 THEN 1110
1050   PRINT N$", HOW MUCH DO YOU WANT TO BET ";
1060   INPUT B[1]
1070   IF B[1] <= (M[1] MIN M[2]) MIN 500 THEN 1140
1080   PRINT "ENTER AN AMOUNT LESS THAN OR EQUAL TO ";
1090   PRINT (M[1] MIN M[2]) MIN 500;
1100   GOTO 1060
1110   Q=1 MAX (10*INT(M[2]/150-M[2]*RND(1)/300))
1120   B[2]=Q MIN (500 MIN M[1])
1130   PRINT "JASON BETS "B[2]
1140   PRINT LIN(1)
1150   IF T=2 THEN 2380
1155   REM CARDS NUMBER P+1, P+3 ARE THE PLAYER'S DOWN CARDS,
1156   REM CARD NUMBER P+2 IS THE DEALER'S UP-CARD, AND CARD
1157   REM NUMBER P+4 IS THE DEALER'S DOWN CARD
1160   PRINT N$",YOUR HOLE CARDS ARE"
1170   PRINT
1180   GOSUB 3660
1190   GOSUB 3700
1200   PRINT
1210   PRINT "AND "D$"'S UP-CARD IS ";TAB(36);
1220   J=P+2
1230   GOSUB 3620
1240   GOSUB 3840
1250   IF H1=21 OR G1=21 THEN 1940
1260   PRINT LIN(1)
1270   IF G[P+1] <> G[P+3] THEN 1490
1275   REM IF THE OPERATOR IS THE PLAYER AND HE HAS A PAIR,
1276   REM HE IS ASKED IF HE WANTS TO SPLIT HIS HAND
1280   PRINT N$" ,DO YOU WANT TO SPLIT YOUR HAND ";
1290   INPUT A$
1300   IF A$[1,1]="N" THEN 1490
1310   S=1
1320   GOTO 1350
1330   PRINT
1340   S=2
1350   PRINT "HERE IS THE"
1360   J=P+N+2
1370   G[P+3]=G[J]
1380   GOSUB 3700
1390   GOSUB 3620
1400   PRINT "TO YOUR"
1410   GOTO S OF 1420,1440
1420   J=P+1
1430   GOTO 1450
1440   J=P+3
1450   GOSUB 3620
1460   N=N-1
1470   IF G[P+1] <> 1 THEN 1590
1480   GOTO S OF 1740,1770
1490   GOSUB 3700
1500   IF G0=G1 AND G1 >= 9 AND G1 <= 11 THEN 1540
1510   IF G0 <> G1 AND G1 >= 13 AND G1 <= 18 THEN 1540
1520   IF G1=8 AND G[P+2]=5 AND (G[P+1]=3 OR G[P+3]=3) THEN 1540
```

*Fig. 5.3 Master program "Blackjack" (cont'd)*

```
1530      GOTO 1590
1535      REM IF THE CONDITIONS IN LINES 1500 TO 1520 ARE MET,
1536      REM THE PLAYER IS ASKED IF HE WANTS TO DOUBLE DOWN
1540      PRINT "DO YOU WANT TO DOUBLE DOWN ";
1550      INPUT A$
1560      IF A$[1,1]="N" THEN 1590
1570      B[T]=2*B[T]
1580      D=1
1590      GOTO D+1 OF 1600,1630
1600      PRINT "DO YOU WANT A HIT ";
1610      INPUT A$
1620      IF A$[1,1]="N" THEN 1710
1630      PRINT "HERE IS THE"
1640      J=P+N+2
1650      GOSUB 3760
1660      GOSUB 3620
1665      REM THE NUMBER OF THE FIRST CARD THE DEALER IS TO DRAW
1666      REM IS FROZEN AND DENOTED BY M
1670      M=P+N+2
1680      IF G1>21 AND S=0 THEN 2090
1690      IF G1>21 THEN 1710
1700      GOTO D+1 OF 1600,1760
1710      PRINT LIN(1)
1715      REM SEE THE REMARK IN LINES 1665,1666
1720      M=N+P+2
1730      IF S <> 1 THEN 1760
1735      REM G IS THE COUNT OF THE FIRST HAND OF A SPLIT HAND
1740      G=G1
1750      GOTO 1330
1760      IF S=0 THEN 1800
1770      IF G <= 21 OR G1 <= 21 THEN 1800
1775      REM N IS ADVANCED BY 1 BECAUSE BOTH OF THE PLAYER'S
1776      REM HANDS (OBTAINED BY SPLITTING) ARE BUSTED AND THE
1777      REM DEALER'S DOWN CARD IS BURNED WITHOUT BEING
1778      REM DISPLAYED
1780      N=N+1
1790      GOTO 2090
1800      PRINT
1810      PRINT D$"'S HOLE CARD IS ";TAB(36);
1820      J=P+4
1830      GOSUB 3620
1840      IF H1>21 THEN 2150
1850      IF H1<17 THEN 1890
1860      PRINT LIN(1)
1870      GOTO 2+SGN(21-G1) OF 2090,1880,1880
1880      GOTO SGN(H1-G1)+2 OF 2000,2030,2060
1890      PRINT D$" DRAWS THE ";TAB(36);
1900      J=P+N+1
1910      GOSUB 3920
1920      GOSUB 3620
1930      GOTO 1840
1940      GOSUB 4030
1950      J=P+4
1960      PRINT D$"'S HOLE CARD IS ";TAB(36);
1970      GOSUB 3620
1980      PRINT LIN(1)
1990      GOTO 1880
2000      PRINT D$" LOSES ";
2010      C[S+1]=B[T]+(W-1)*B[T]/2
2020      GOTO 2270
2030      PRINT D$" PUSHES ";
```

*Fig. 5.3 Master program "Blackjack" (cont'd)*

```
2040   C[S+1]=0
2050   GOTO 2270
2060   PRINT D$" WINS ";
2070   C[S+1]=-B[T]
2080   GOTO 2270
2090   PRINT LIN(1)
2100   PRINT P$" IS BUSTED WITH A COUNT OF "G1" ";
2110   C[S+1]=-B[T]
2120   IF S <> 0 THEN 2300
2125   REM N IS ADVANCED BY 1 BECAUSE THE PLAYER'S (OPERATOR'S)
2126   REM HAND IS BUSTED AND THE DEALER'S (COMPUTER'S) DOWN
2127   REM CARD IS BURNED WITHOUT BEING DISPLAYED. NOTE THAT
2128   REM NOTHING HAPPENS IF T=2, I.E., THE COMPUTER IS THE
2129   REM PLAYER
2130   N=N+(2-T)
2140   GOTO 730
2150   PRINT LIN(1)
2155   REM LINES 2160 TO 2370 TAKE CARE OF ANNOUNCING THE
2156   REM RESULT OF THE COUP
2160   PRINT D$" IS BUSTED WITH A COUNT OF "H1" ";
2170   C[S+1]=B[T]
2180   IF S <> 2 THEN 730
2190   C[S+1]=2*B[T]
2200   S=1
2210   IF G <= 21 AND G1 <= 21 THEN 730
2220   PRINT "BUT "P$" IS BUSTED WITH A COUNT OF "G MAX G1" ";
2230   C[2]=-2*B[T]
2240   IF G>21 THEN 2360
2250   PRINT "ON THE SECOND HAND."
2260   GOTO 730
2270   PRINT "WITH A COUNT OF "H1" VS. "P$"'S COUNT OF "G1" ";
2280   IF S <> 0 THEN 2300
2290   GOTO 730
2300   IF H=1 THEN 2360
2310   PRINT "ON THE SECOND HAND AND ";
2320   S=1
2330   G1=G
2335   REM H, NORMALLY 0, IS SET TO 1 TO REGULATE IN LINE 2300
2336   REM THE ANNOUNCEMENT OF THE RESULTS FROM PLAYING A SPLIT
2337   REM HAND
2340   H=1
2350   GOTO 1860
2360   PRINT "ON THE FIRST HAND."
2370   GOTO 730
2380   PRINT N$", YOUR UP-CARD IS ";TAB(36);
2390   J=P+2
2400   GOSUB 3620
2410   PRINT "AND YOUR HOLE CARD IS ";TAB(36);
2420   J=P+4
2430   GOSUB 3620
2440   PRINT
2450   GOSUB 3840
2460   GOSUB 3700
2470   IF G1 <> 21 AND H1 <> 21 THEN 2540
2480   GOSUB 4030
2490   PRINT P$"'S HOLE CARDS ARE"
2500   PRINT
2510   GOSUB 3660
2520   PRINT
2530   GOTO 1880
2535   REM LINES 2540 TO 2900 EMBODY THE OPTIMAL (ZERO MEMORY)
```

*Fig. 5.3 Master program "Blackjack" (cont'd)*

```
2536   REM STRATEGY TO BE FOLLOWED BY THE COMPUTER AS PLAYER
2537   REM LINES 2540 TO 2640 TAKE CARE OF PAIRS
2540   IF G[P+1] <> G[P+3] THEN 2650
2550   IF G[P+1]=1 THEN 2630
2560   IF G[P+1]=10 OR G[P+1]=5 THEN 2660
2570   IF G[P+1]=4 AND G[P+2] <> 5 THEN 2660
2580   IF G[P+1] <= 6 AND (G[P+2] >= 8 OR G[P+2]=1) THEN 2660
2590   IF G[P+1]=7 AND (G[P+2] >= 9 OR G[P+2]=1) THEN 2660
2600   IF G[P+1] <> 9 THEN 2630
2610   IF G[P+2] <> 7 AND G[P+2] <> 10 AND G[P+2] <> 1 THEN 2630
2620   GOTO 2660
2630   PRINT P$" IS GOING TO SPLIT HIS HAND."
2640   GOTO 3400
2645   REM LINES 2650 TO 2730 TAKE CARE OF EXCEPTIONAL HANDS
2650   IF G1 <> G0 THEN 2820
2660   IF G1=13 AND G[P+2]=2 AND (G[P+1]=3 OR G[P+3]=3) THEN 2910
2670   IF G1=12 AND G[P+2]=4 AND (G[P+1]=2 OR G[P+3]=2) THEN 2910
2680   IF G1 <> 12 OR G[P+2] <> 3 THEN 2710
2690   IF G[P+1] >= 3 AND G[P+1] <= 5 THEN 3160
2700   IF G[P+3] >= 3 AND G[P+3] <= 5 THEN 3160
2710   IF G1=8 AND G[P+2]=6 AND (G[P+1]=2 OR G[P+3]=2) THEN 3160
2720   IF G1=8 AND G[P+2]=5 AND (G[P+1]=3 OR G[P+3]=3) THEN 3070
2730   IF G1=14 AND G[P+2]=10 AND G[P+1]=7 THEN 3160
2735   REM LINES 2740 TO 2810 TAKE CARE OF HARD HANDS
2740   IF G1 >= 17 THEN 3160
2750   IF G1>12 AND G[P+2]>1 AND G[P+2]<7 THEN 3160
2760   IF G1=12 AND G[P+2]>3 AND G[P+2]<7 THEN 3160
2770   IF G1=11 THEN 3070
2780   IF G1=10 AND G[P+2] <> 10 AND G[P+2] <> 1 THEN 3070
2790   IF G1=9 AND G[P+2]>1 AND G[P+2]<7 THEN 3070
2800   IF G1=8 AND G[P+2]=6 THEN 3070
2810   GOTO 2910
2815   REM LINES 1820 TO 2900 TAKE CARE OF SOFT HANDS
2820   IF G1>18 THEN 3160
2830   IF G1 <> 18 THEN 2870
2840   IF G[P+2]>2 AND G[P+2]<7 THEN 3070
2850   IF G[P+2]>8 THEN 2910
2860   GOTO 3160
2870   IF G1=17 AND G[P+2]>1 AND G[P+2]<7 THEN 3070
2880   IF G1=17 THEN 2910
2890   IF G[P+2]<4 OR G[P+2]>6 THEN 2910
2900   GOTO 3070
2910   PRINT P$" DRAWS THE"
2915   REM IN LINE 2920, THE PLAYER, PLAYING A STRAIGHT HAND,
2916   REM DRAWS THE 5-TH CARD OF THE DEAL ON THE FIRST DRAW,
2917   REM THE 6-TH CARD OF THE DEAL IF HE DRAWS TO THE FIRST
2918   REM HAND OF A SPLIT HAND AFTER THE HIT, ETC...
2920   J=N+P+3-S
2930   GOSUB 3620
2940   GOSUB 3760
2945   REM THE SIMPLE STRATEGY (SS) FOR ADDITIONAL DRAWS IS
2946   REM EMBODIED IN LINES 2950 TO 3060
2950   IF G0 <> G1 THEN 3030
2960   IF G1<12 OR G1>16 THEN 2990
2970   IF G[P+2] <> 1 AND (G[P+2]<7 OR G[P+2]>10) THEN 2990
2980   GOTO 3050
2990   IF G1<12 THEN 3050
3000   IF S <> 0 THEN 3200
3010   PRINT "TO THE"
3020   GOTO 3190
3030   IF G1<18 THEN 3050
```

Fig. 5.3 Master program "Blackjack" (cont'd)

```
3040    GOTO 3000
3050    PRINT "AND THE"
3060    GOTO 2920
3065    REM LINE 3070 PREVENTS DOUBLING DOWN ON A SPLIT HAND
3070    GOTO S+1 OF 3080,2910,2910
3080    PRINT P$" DOUBLES DOWN WITH THE HOLE CARDS"
3090    B[T]=2*B[T]
3100    GOSUB 3660
3110    PRINT "AND TAKES HIS OBLIGATORY HIT:"
3120    J=P+N+1
3130    GOSUB 3620
3140    GOSUB 3760
3150    GOTO 3240
3160    PRINT LIN(1)
3170    GOTO S+1 OF 3180,3420,3240
3180    PRINT P$" STAYS WITH THE HOLE CARDS"
3190    GOSUB 3660
3200    GOTO S OF 3210,3230
3210    G=G1
3220    IF S=1 THEN 3420
3230    IF G1>21 AND G>21 THEN 2090
3240    PRINT LIN(1)
3250    IF H1 >= 17 THEN 1860
3260    PRINT D$", DO YOU WANT A HIT ";
3270    M=P+N+1
3280    INPUT A$
3290    IF A$[1,1]="N" THEN 1860
3300    PRINT LIN(1)
3310    PRINT "HERE IS THE "TAB(36);
3320    J=P+N+1
3330    GOSUB 3620
3340    GOSUB 3920
3350    IF H1 <= 21 THEN 3370
3360    GOTO 2150
3365    REM IF THE COUNT OF THE COMPUTER'S HAND, THE COMPUTER
3366    REM BEING THE DEALER, EXCEEDS 16, CONTROL IS PASSED
3367    REM TO THE WIND-UP ROUTINE STARTING WITH LINE 1860
3370    IF H1 >= 17 THEN 1860
3380    PRINT "DO YOU WANT A HIT ";
3390    GOTO 3280
3395    REM LINES 3400 TO 3610 DEAL WITH A SPLIT HAND IF THE
3396    REM COMPUTER IS THE PLAYER
3400    S=1
3410    GOTO 3440
3420    PRINT
3430    S=2
3440    PRINT "HERE IS THE"
3450    GOTO S OF 3460,3480
3460    J=N+P+3
3470    GOTO 3490
3480    J=P+N+2
3490    G[P+3]=G[J]
3500    GOSUB 3700
3510    GOSUB 3620
3520    PRINT "TO "P$"'S"
3530    GOTO S OF 3540,3560
3540    J=P+1
3550    GOTO 3570
3560    J=P+3
3570    GOSUB 3620
3580    GOTO S OF 3590,3600
```

Fig. 5.3 Master program "Blackjack" (cont'd)

```
3590    G=G1
3600    IF G[P+1] <> 1 THEN 2650
3610    GOTO S OF 3420,3230
3615    REM IN LINES 3620 TO 3650, A CARD IS DISPLAYED AND
3616    REM COUNTED BY ADVANCING N BY ONE UNIT
3620    PRINT "*** "V$[1+5*(V[J]-1),5+5*(V[J]-1)]" OF ";
3630    PRINT S$[1+8*L[J],8+8*L[J]]" ***"
3640    N=N+1
3650    RETURN
3655    REM IN THE SUBROUTINE 3660 TO 3690, CARDS NUMBER
3656    REM P+1, P+3 ARE DEALT TO THE PLAYER
3660    FOR J=P+1 TO P+3 STEP 2
3670    GOSUB 3620
3680    NEXT J
3690    RETURN
3695    REM IN THE SUBROUTINE 3700 TO 3750, THE PLAYER'S
3696    REM DOWN CARDS ARE EVALUATED
3700    G0=G[P+1]+G[P+3]
3710    IF G[P+1] <> 1 AND G[P+3] <> 1 THEN 3740
3720    G1=G0+10
3730    RETURN
3740    G1=G0
3750    RETURN
3755    REM IN THE SUBROUTINE 3760 TO 3830, THE PLAYER'S HAND
3756    REM IS EVALUATED AFTER HE HAS DRAWN CARD NUMBER J
3760    IF G0=G1 AND G[J]=1 AND G1+G[J]+10 <= 21 THEN 3790
3770    G1=G1+G[J]
3780    GOTO 3800
3790    G1=G1+G[J]+10
3800    G0=G0+G[J]
3810    IF G1 <= 21 THEN 3830
3820    G1=G0
3830    RETURN
3835    REM IN THE SUBROUTINE 3840 TO 3910, THE DEALER'S FIRST
3836    REM TWO CARDS ARE EVALUATED
3840    H0=G[P+2]+G[P+4]
3850    IF G[P+2] <> 1 AND G[P+4] <> 1 THEN 3880
3860    IF H0<7 OR H0>11 THEN 3880
3870    GOTO 3900
3880    H1=H0
3890    GOTO 3910
3900    H1=H0+10
3910    RETURN
3915    REM IN THE SUBROUTINE 3920 TO 4020, THE DEALER'S HAND
3916    REM IS EVALUATED AFTER HE HAS DRAWN CARD NUMBER J
3920    H0=H0+G[J]
3930    IF G[P+2]=1 OR G[P+4]=1 THEN 3980
3940    FOR I=M TO J
3950    IF G[I]=1 THEN 3980
3960    NEXT I
3970    GOTO 3990
3980    IF H0 >= 7 AND H0 <= 11 THEN 4010
3990    H1=H0
4000    RETURN
4010    H1=H0+10
4020    RETURN
4030    PRINT LIN(1)
4040    PRINT TAB(15)"********** BLACKJACK **********"
4045    REM IF THE PLAYER HAS A BLACKJACK, W IS SET TO 2 IN
4046    REM LINE 4050 SINCE SGN(21-H1)=1
4050    W=1+SGN(21-H1)
4060    PRINT LIN(1)
```

*Fig. 5.3 Master program "Blackjack" (cont'd)*

```
4070    RETURN
4075    REM IN LINES 4080 TO 4340, 4 DECKS OF CARDS ARE
4076    REM SHUFFLED AND CUT
4080    PRINT TAB(7)"*** STAND BY - ";
4090    PRINT "THE CARDS ARE BEING SHUFFLED ***"
4100    PRINT LIN(1)
4110    FOR W=0 TO 3
4120    FOR J=1+52*W TO 52+52*W
4130    S[J]=J
4140    NEXT J
4150    FOR I=52+52*W TO 2 STEP -1
4160    J=INT(RND(1)*I)+1
4170    Z=S[J]
4180    S[J]=S[I]
4190    S[I]=Z
4200    NEXT I
4210    NEXT W
4220    FOR I=208 TO 2 STEP -1
4230    J=INT(RND(1)*I)+1
4240    Z=S[J]
4250    S[J]=S[I]
4260    S[I]=Z
4270    NEXT I
4280    PRINT TAB(1)"*** PLEASE CUT - BY ENTERING A NUMBER ";
4290    PRINT "BETWEEN 1 AND 208 ***"
4300    PRINT LIN(1)
4310    INPUT Z
4320    FOR J=1 TO 208
4330    T[J]=S[J]+Z-208*INT((S[J]+Z-1)/208)
4340    NEXT J
4345    REM IN LINES 4350 TO 4400, VALUE V(J), SUIT S(J),
4346    REM AND GAME VALUE G(J) OF CARD NUMBER S(J) IS
4347    REM EVALUATED FOR ALL J FROM 1 TO 208
4350    FOR J=1 TO 208
4360    S[J]=T[J]-52*INT((T[J]-1)/52)
4370    V[J]=S[J]-13*INT((S[J]-1)/13)
4380    L[J]=INT((S[J]-1)/13)
4390    G[J]=V[J] MIN 10
4400    NEXT J
4410    P=0
4420    RETURN
4430    PRINT LIN(1)
4440    IF M[1] <= 0 THEN 4470
4450    PRINT "CONGRATULATIONS ! YOU WHIPPED JASON REAL GOOD !"
4460    GOTO 4480
4470    PRINT "JASON SURE SLAUGHTERED YOU !"
4480    END
```

Fig. 5.3 Master program "Blackjack" (cont'd)

to keep the bookkeeping straight, and, using the computed GOTO statement to advantage, you'll have to make provisions for keeping track of the counts of the various hands and also for the computer's final announcement of the outcome of the coup. We do not think it worthwhile to make provisions for the splitting of a hand that has been obtained from splitting a split hand.

To enable a player to double down on a split hand, you must adapt the double-down routine to our splitting routine. Don't forget that a player who has doubled down gets exactly one hit per hand.

It is a relatively simple matter to let a player take out insurance if the dealer's up card is an ace [G(P+2)=1]. In such a case, if the dealer does not have a blackjack, the player loses B(T)/2 in addition to losing B(T), if he loses, or winning B(T), if he wins. He breaks even if the dealer does have a blackjack.

For a friendly game, with the deal passing back and forth between you and "Jason" (the computer), you may wish to change the pay-off on a blackjack from 3:2 to 2:1. This is easy:

    2010   C(S+1)=B(T)+(W−1)∗B

If you want to play casino-style with the computer as the permanent dealer, all you need to add to the program in Fig. 5.3 is the following:

              450   GOTO 600
              860   GOTO 880

This will give Jason the first deal, and nothing will ever let it slip out of his hand.

You may wish to use this modified program to learn to play blackjack. In the beginning, play with the aid of Tables 5.1 through 5.4; as soon as you think that you have the hang of it, dispense with tables.

If you have studied some or all of Thorp's counting methods, you may use this modified program to perfect your game. Suppose you want to use his "counting fives" method. You may program the computer to let you know when all the 5's have been revealed. (This is not the same as all the 5's having been dealt. If the player's hand is busted, the dealer's down card is burned without being revealed, and hence some 5's may slip by the count!) Noting that there are 16 5's in four decks, you may go about this task as follows:

```
4411   F=0
3611   IF G[J] <> 5 THEN 3620
3612   F=F+1

3641   GOTO SGN(ABS(F-16))+1 OF 3642,3650
3642   PRINT TAB(14)"SWITCH TO THORP ! SWITCH TO THORP !"
3643   PRINT TAB(14)"*****    ALL FIVES ARE GONE    *****"
3644   F=100
```

Change the GOSUB 3620 in lines 510, 550, 1230, 1390, 1660, 1830, 1920, 1970, 2400, 2430, 2930, 3130, 3330, 3510, and 3670 to

                     GOSUB 3611

(that is, in all cases except in line 1450, where such a change would subject two 5's to being counted twice if the player's hand holds two 5's and the player is the operator. They would be counted the first time when they are dealt and a second time when each of them is hit and the card being hit is mentioned again.)

As soon as you are informed that all 5's have been revealed (and are, in fact, gone), you switch your strategy to Thorp's "Best Strategy When It Is Only Known That No Fives Can Appear on the Next Round of Play" (Ref. [13], p. 49).

The program's bookkeeping capability will let you know how well you are holding your own against the house. If you keep winning consistently, you may be ready for the real thing! Of course, you will have to maintain your attitude towards money. Either you treat the play money on the computer as you would treat real money, or else, you'll have to play with your real money as if it were play money.

For the case where the deal passes back and forth, you may wish the computer to use the "counting fives" strategy when it is the player. As long as there are 5's in the deck, you let the computer play the optimal (zero memory) strategy as before (lines 2540 to 2900). But, as soon as F turns 16, you make the computer switch to Thorp's strategy by means of

```
2540   GOTO SGN(ABS(F-100))+1 OF 4490,2541
2541   IF G(P + 1) <> G(P + 3) THEN 2650
4460   GOTO 4720
4480   GOTO 4720
```

and program Thorp's strategy (Ref. [13], p. 49), beginning with line 4490, as follows:

```
4490   IF GCP+1] <> GCP+3] THEN 4570
4500   IF GCP+1]=1 THEN 2630
4510   IF GCP+1]<4 AND (GCP+2]>8 OR GCP+2]=1) THEN 4570
4520   IF GCP+1]=6 AND (GCP+2]>6 OR GCP+2]=1) THEN 4570
4530   IF GCP+2]=10 AND (GCP+1]=7 OR GCP+1]=9) THEN 4570
4540   IF GCP+2]=1 AND (GCP+1]=7 OR GCP+1]=9) THEN 4570
4550   IF (GCP+1]=10 AND GCP+2] <> 6) OR GCP+1]=4 THEN 4570
4560   GOTO 2630
4570   IF G0 <> G1 THEN 4650
4580   IF G1=10 OR G1=11 THEN 3070
4590   IF G1=9 AND GCP+2]>1 AND GCP+2]<8 THEN 3070
4600   IF G1=8 AND GCP+2]>3 AND GCP+2]<7 THEN 3070
4610   IF G1>11 AND GCP+2]>1 AND GCP+2]<7 THEN 3160
4620   IF G1>16 AND (GCP+2]=7 OR GCP+2]=8 OR GCP+2]=1) THEN 3160
4630   IF G1>14 AND GCP+2]>8 THEN 3160
4640   GOTO 2910
4650   IF GCP+2]>2 AND GCP+2]<7 AND G1>12 AND G1<20 THEN 3070
4660   IF GCP+2]=6 AND G1=20 THEN 3070
4670   IF GCP+2]=2 AND G1>16 AND G1<19 THEN 3070
4680   IF GCP+2]=7 AND G1=17 THEN 3070
4690   IF G1>18 AND GCP+2]>8 THEN 3160
4700   IF G1>17 AND GCP+2]<9 THEN 3160
4710   GOTO 2910
```

Note that the case of two 5's cannot occur because when this part of the program is activated, all the 5's are already gone.

By adding

4720   END

we have a workable program. (We leave it to the reader to unscramble the above instructions and present the strategy in tables as we have done for the optimal (zero memory) strategy in Tables 5.1 through 5.4.) As before, we use the simple strategy for all draws after the first (crucial) draw.

To win with the "counting fives" strategy, one has to increase one's bet as soon as all the 5's are gone and the new strategy goes into effect. Thorp recommends betting the table minimum (of $1 or whatever) as long as there are 5's in the deck and then making your bets as large as possible as soon as the 5's are gone. Suppose you go all the way to the table maximum of $500. You may then expect to make, on the average, $169.82 per hour, assuming that you play 100 hands per hour (Ref. [13], p. 57.) Such a sudden change in staking may arouse the dealer's suspicion, however, prompting him to reshuffle the deck(s) and putting you, in effect, back to square one. Thorp suggests increasing the size of the bet threefold or fourfold instead. To enable the computer (as player) to do this, you may want to add to line 1110, where Jason's bet is calculated, the following two lines:

1111   GOTO SGN(ABS(F−100))+1 OF 1112,1120
1112   Q=4*Q

Just as it is possible to program the "counting fives" method into the computer, it is also possible to program the "ten-count" strategy, or any other strategy, for that matter, as long as it may be formulated in terms of a finite number of words. We leave this task for the reader to carry out when he has completed his study of the higher strategies.

# Appendix A
# Probability

## 1. THE GEOMETRIC PROGRESSION

The sequence 1, q, $q^2$, $q^3$, ... is a geometric progression. The sum of its first n terms is given by

$$(1.1) \qquad 1 + q + q^2 + \ldots + q^{n-1} = \frac{1 - q^n}{1 - q}, \text{ where } q \neq 1$$

Try it out for several values of q and several integral values of n and you will see that you obtain the right answer every time. Magic? No— just some clever algebra.

If we let n become larger and larger, then

$$(1.2) \qquad 1 + q + q^2 + q^3 + \ldots = \frac{1}{1 - q}, \text{ where } -1 < q < 1$$

You may interpret Formula (1.2) thusly: If there are a great number of terms (if n is very large), then the sum on the left is so close to the number $1/(1 - q)$ on the right that you can't tell the difference. Formula (1.2) is obtained from Formula (1.1) by "letting n tend to infinity." Try it out. Take the $q^n$ on the right side of Formula (1.1), and compute successive powers of q on your computer:

```
10   PRINT "ENTER A NUMBER BETWEEN -1 AND 1 ";
20   INPUT Q
30   FOR N=1 TO 1.E+06
40   PRINT  USING "#,D.6DX";Q^N
50   IF N/7 <> INT(N/7) THEN 70
60   PRINT
70   IF ABS(Q^N)<1.E-07 THEN 90
80   NEXT N
90   END
```

We ran this program several times and here is what happened:

```
RUN


ENTER A NUMBER BETWEEN -1 AND 1 ?.01
0.010000 0.000100 0.000001 0.000000
DONE

RUN


ENTER A NUMBER BETWEEN -1 AND 1 ?.1
0.100000 0.010000 0.001000 0.000100 0.000010 0.000001 0.000000

DONE

RUN


ENTER A NUMBER BETWEEN -1 AND 1 ?.5
0.500000 0.250000 0.125000 0.062500 0.031250 0.015625 0.007813
0.003906 0.001953 0.000977 0.000488 0.000244 0.000122 0.000061
0.000031 0.000015 0.000008 0.000004 0.000002 0.000001 0.000000
0.000000 0.000000 0.000000
DONE

RUN


ENTER A NUMBER BETWEEN -1 AND 1 ?.9
0.900000 0.810000 0.729000 0.656100 0.590490 0.531441 0.478297
0.430467 0.387420 0.348678 0.313810 0.282429 0.254186 0.228768
0.205891 0.185302 0.166772 0.150095 0.135085 0.121577 0.109419
0.098477 0.088629 0.079766 0.071790 0.064611 0.058150 0.052335
0.047101 0.042391 0.038152 0.034337 0.030903 0.027813 0.025032
0.022528 0.020276 0.018248 0.016423 0.014781 0.013303 0.011972
0.010775 0.009698 0.008728 0.007855 0.007070 0.006363 0.005726
0.005154 0.004638 0.004175 0.003757 0.003381 0.003043 0.002739
0.002465 0.002219 0.001997 0.001797 0.001617 0.001456 0.001310
0.001179 0.001061 0.000955 0.000860 0.000774 0.000696 0.000627
0.000564 0.000508 0.000457 0.000411 0.000370 0.000333 0.000300
0.000270 0.000243 0.000218 0.000197 0.000177 0.000159 0.000143
0.000129 0.000116 0.000104 0.000094 0.000085 0.000076 0.000069
0.000062 0.000056 0.000050 0.000045 0.000040 0.000036 0.000033
0.000030 0.000027 0.000024 0.000022 0.000019 0.000017 0.000016
0.000014 0.000013 0.000011 0.000010 0.000009 0.000008 0.000008
0.000007 0.000006 0.000005 0.000005 0.000004 0.000004 0.000004
0.000003 0.000003 0.000003 0.000002 0.000002 0.000002 0.000002
0.000002 0.000001 0.000001 0.000001 0.000001 0.000001 0.000001
0.000001 0.000001 0.000001 0.000001 0.000000 0.000000 0.000000
0.000000 0.000000 0.000000 0.000000 0.000000 0.000000 0.000000
0.000000 0.000000 0.000000 0.000000 0.000000 0.000000
DONE
```

Does that convince you? No? Well, then you'll just have to study some mathematics.

The left side of Formula (1.2) is called the *geometric series* and the expression on the right is called the *sum of the geometric series*. Needless to say, it is meaningless when $q \leq -1$ or $q \geq 1$.

## 2. PROBABILITY

The ancient Greek Philosopher Aristotle[1] once said, "Probable is what is most likely to happen." (He said it in Greek, which may have endeared him to his contemporaries, but it certainly does not endear him to our present-day, hemi-literate student bodies and faculties.) Now, what is most likely to happen? Presumably, what's probable—and here is a classical (no pun intended) case of a circular definition. Pierre Simon, the Marquis de Laplace,[2] French mathematician, physicist, and, for a mercifully brief period of time, a member of Napoleon Bonaparte's cabinet as minister of the interior, did not do much better. He said that among a given number of equally likely events, the probability that a favorable event will occur is given by the number of favorable events, divided by the number of possible events:

$$(2.1) \qquad \text{Probability} = \frac{\text{Number of favorable events}}{\text{Number of possible events}}$$

What are "equally likely" events? Presumably, events that occur with equal probability—and the vicious circle is closed!

Still, Formula (2.1) provides us with a useful working definition even though it is neither logically acceptable nor intellectually satisfying. (The alternative would be unacceptable for the very practical reason that it would lead us too far afield.)

Here we go: If one spins a European roulette wheel, there are 37 equally likely outcomes: 0, 1, 2, 3, . . . , 36. Eighteen of them are rouge (red) and 18 are noir (black); 18 are pair (even) and 18 are impair (odd); 18 are passe (high) and 18 are manque (low). Hence, by Formula (2.1),

$$\text{Probability of rouge} = 18/37 = 0.4864864$$

and the same probability is obtained for noir, pair, impair, passe, and manque. On the other hand,

$$\text{Probability of zero} = 1/37 = 0.027027$$

In trente-et-quarante, out of 6589 (equally likely) valid coups, 169 result in a refait,[3] and hence,

$$\text{Probability of refait} = 169/6589 = 0.0256488$$

while half the valid coups that do not result in a refait—namely, half of 6420 (6589 − 169), or 3210—result in a noir (or rouge, couleur, or inverse):

---

[1] 384–322 B.C.
[2] 1749–1827
[3] See section 5 of this Appendix.

Probability of *noir* = 3210/6589 = 0.4871755

Get the idea?

From Formula (2.1), we obtain the following:

$$\text{Probability of certain event} = \frac{\text{Number of possible events}}{\text{Number of possible events}} = 1$$

and

$$\text{Probability of impossible event} = \frac{0}{\text{Number of possible events}} = 0$$

## 3. ODDS

If an event, call it E, occurs with probability p, then its opposite will occur with probability $1 - p$. The ratio,

$$p:(1 - p)$$

is called the *odds in favor of the event E*.

For example, the odds for *rouge* in European roulette are

$$18/37:(1 - 18/37) = 18:(37 - 18) = 18:19$$

and the odds in favor of a *refait* in trente-et-quarante are

$$169/6589:(1 - 169/6589) = 169:(6589 - 169) = 169:6420$$

Conversely, if the odds in favor of an event E are given to be

$$a:b$$

then the probability that E will occur is given by

$$p = \frac{a}{a + b}$$

[Proof: $p/(1 - p) = (a/(a + b))/(1 - (a/(a + b))) = (a/(a + b))/(b/(a + b)) = a/b$]

## 4. THE PROBABILITY TREE AND TREE PROBABILITIES

We flip a coin until it comes up heads and then jump in the lake. What is the probability that heads will come up for the first time at the first flip, the second flip, the third flip, etc.? Let's look at the self-explanatory diagram in Fig. A.1. We call this a *probability tree*. Heads and tails are two equally likely events. Hence, the probability for the favorable event heads is

Probability of heads = 1/2

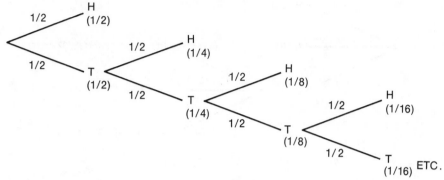

Fig. A.1 Probability tree

Consequently, the probability of heads at the first throw is 1/2. If heads shows, we jump in the lake, and that's the end of it. But suppose that tails comes up on the first throw. After that, heads and tails may occur with probability 1/2 each. So, out of four possible events—HH, HT, TH, and TT—we have one favorable event, namely TH. Hence, we obtain for the probability of a tails followed by a heads the value 1/2 × 1/2, or 1/4. Similarly, we obtain 1/8 for the probability of two tails in a row to be followed by a heads, 1/16 for the probability of three tails to be followed by a heads, etc. We call these probabilities the *tree probabilities*. They are computed by taking the product of the probabilities of the individual branches that lead up to the event we are interested in—in this case, heads. The probability that heads will come up eventually is obtained from taking the sum of the probabilities that heads will come up the first time, plus the probability that heads will come up the second time only, plus the probability that it will come up the third time only, etc. By Formula (1.2):

$$\frac{1}{2} + \left(\frac{1}{2}\right)^2 + \left(\frac{1}{2}\right)^3 + \ldots = \frac{1}{2}\left[1 + \frac{1}{2} + \left(\frac{1}{2}\right)^2 + \ldots\right] = \frac{1}{2}\left[1 \middle/ \left(1 - \frac{1}{2}\right)\right] = 1$$

that is, heads will have to come up eventually!

Let's play European roulette with one degree of prison and *no* "half-your-stake-back" option after a zero. Suppose we back *rouge*. The probability tree is shown in Fig. A.2.

We lose (1) if the first spin results in a *noir* or (2) if the first spin results in a zero and the second spin results in a *noir* or a zero. Hence, the (tree) probability of losing is given by

Probability of losing a bet on *rouge* = 18/37 + (1/37)(18/37) + (1/37)²
$$= 685/1369 = 0.5003652$$

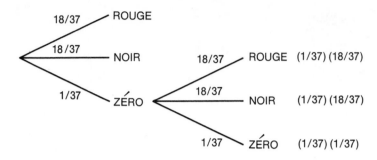

Fig. A.2 Probability tree for European roulette

Similarly, we obtain the probability of winning as follows:

Probability of winning a bet on *rouge* = 18/37 = 0.4864864

and the probability of breaking even (a zero followed by a *rouge*) as follows:

Probability of breaking even = (1/37)(18/37) = 0.0131482

These three probabilities should add up to 1 (one of the three events has to occur for sure), and they will if you add up the fractions. They don't quite do so if you add up the seven-place decimal approximations; how could they? (By using the appropriate probability tree, the reader can show that, allowing prisons of higher degree, the probability of losing actually decreases, although by minute amounts. As a matter of fact, allowing prisons of second degree, the probability of losing is 0.500190 rather than 0.5003652, a decrease of 0.035 percent! Allowing prisons of third degree, the probability decreases to 0.500188. The tree method is too clumsy to deal with prisons of third and higher degrees efficiently. More sophisticated methods, such as those involving matrix inversion, are recommended (Ref. [6], p. 214ff, and Ref. [9]).

## 5. SOME SIMPLE COMBINATORICS

With two dice, one may roll any number between 2 and 12. What is the probability of rolling, say, a 7? Well, how many (equally likely) favorable events are there and how many (equally likely) possible events? Table A.1· is a self-explanatory table that shows all the combinations resulting in a roll of 2 to 12 points.

We see that there are 36 possible events, and only one of them results in a 2, but two yield a 3, three yield a 4, . . . , six yield a 7, five yield an 8, etc. If the dice are perfect (which they are only if all 36 events are equally likely, of course!), we may assume that all 36 events are equally likely and we obtain

**Table A.1**

| Die No. 2 | Die No. 1 | | | | | |
|---|---|---|---|---|---|---|
| | 1 | 2 | 3 | 4 | 5 | 6 |
| 1 | 2 | 3 | 4 | 5 | 6 | 7 |
| 2 | 3 | 4 | 5 | 6 | 7 | 8 |
| 3 | 4 | 5 | 6 | 7 | 8 | 9 |
| 4 | 5 | 6 | 7 | 8 | 9 | 10 |
| 5 | 6 | 7 | 8 | 9 | 10 | 11 |
| 6 | 7 | 8 | 9 | 10 | 11 | 12 |

$$\text{Probability of rolling 2} = \text{Probability of rolling 12} = \frac{1}{36} = 0.0277$$

$$\text{Probability of rolling 3} = \text{Probability of rolling 11} = \frac{2}{36} = 0.05555$$

$$\text{Probability of rolling 4} = \text{Probability of rolling 10} = \frac{3}{36} = 0.08333$$

$$\text{Probability of rolling 5} = \text{Probability of rolling 9} = \frac{4}{36} = 0.11111$$

$$\text{Probability of rolling 6} = \text{Probability of rolling 8} = \frac{5}{36} = 0.13888$$

$$\text{Probability of rolling 7} = \frac{6}{36} = 0.16666$$

A come-out roll in craps "wins" if the shooter rolls a 7 or 11. Since 8 out of 36 possible outcomes (2 + 6) produce the "favorable event" 7 or 11, we have

$$\text{Probability of 7 or 11} = \frac{8}{36} = \frac{2}{9} = 0.2222222$$

A come-out roll "loses" if the shooter rolls a 2, 3 or 12. Such a result is produced in 4 out of 36 possible cases (1 + 2 + 1). Hence,

$$\text{Probability of 2, 3, or 12} = \frac{4}{36} = \frac{1}{9} = 0.1111111$$

In the remaining 24 cases (36 − 8 − 4), a point (4, 5, 6, 8, 9, or 10) is established, and we have

$$\text{Probability of 4, 5, 6, 8, 9, or 10} = \frac{24}{36} = \frac{2}{3} = 0.6666666$$

Note that if a point is established, then, all subsequent rolls that produce neither the point nor a 7 are irrelevant. If the point is a 4, for

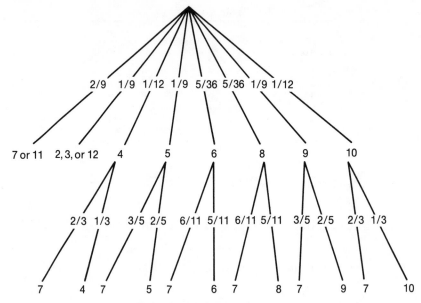

Fig. A.3 Probability tree for craps

example, there are only 9 possible events (3 + 6) from then on, namely a 4 or 7. Of these 9 possible events, only 3 are favorable, etc.

From the probability tree in Fig. A.3 for a series of rolls from the come-out roll to termination (come-out roll wins, come-out roll loses, or point is established and made or the shooter sevens out) we obtain the probability of "winning a pass" as follows:

$$\text{Probability of winning a pass} = \frac{2}{9} + \frac{1}{12} \cdot \frac{1}{3} + \frac{1}{9} \cdot \frac{2}{5} + \frac{5}{36} \cdot \frac{5}{11} + \frac{5}{36} \cdot \frac{5}{11}$$

$$+ \frac{1}{9} \cdot \frac{2}{5} + \frac{1}{12} \cdot \frac{1}{3} = \frac{488}{990} = \frac{244}{495} = 0.49292929$$

Next, let's have a look at trente-et-quarante. We assume that any one of the 13 different cards (the suit does not matter) is equally likely to turn up as the last card in a row. If that card is an ace, then the count must be 31. If the count were anything else, the last card couldn't be an ace. If it is a deuce, the count must be 31 or 32; if it is a trey, the count must be 31, 32, or 33; if it is a 4, the count must be 31, 32, 33, or 34; etc. Finally, if the last card is a 10 or a face card, then the count must be 31, 32, 33, 34, . . . , 39, or 40. Hence, there are

4 ways of getting a count of 40 (10,J,Q,K)
5 ways of getting a count of 39 (9,10,J,Q,K)
6 ways of getting a count of 38 (8,9,10,J,Q,K)

7 ways of getting a count of 37 (7,8,9,10,J,Q,K)
8 ways of getting a count of 36 (6,7,8,9,10,J,Q,K)
9 ways of getting a count of 35 (5,6,7,8,9,10,J,Q,K)
10 ways of getting a count of 34 (4,5,6,7,8,9,10,J,Q,K)
11 ways of getting a count of 33 (3,4,5,6,7,8,9,10,J,Q,K)
12 ways of getting a count of 32 (2,3,4,5,6,7,8,9,10,J,Q,K)
13 ways of getting a count of 31 (Ace,2,3,4,5,6,7,8,9,10,J,Q,K)

Since $4 + 5 + 6 + 7 + 8 + 9 + 10 + 11 + 12 + 13 = 85$, we see that out of 85 possible events, there are 4, 5, 6, . . . , or 13 favorable events. Therefore, using the notation

$$P(n) = \text{Probability of getting a count of } n$$

we obtain

$$P(31) = 13/85, \; P(32) = 12/85, \; P(33) = 11/85, \; P(34) = 10/85,$$
$$P(35) = 9/85, \; P(36) = 8/85, \; P(37) = 7/85, \; P(38) = 6/85,$$
$$P(39) = 5/85, \; P(40) = 4/85$$

The probability tree in Fig. A.4 (which we have drawn only partially because of obvious space limitations) will enable us to figure the probabilities of the various *coups*.

We see from the tree probabilities that $P(n,m)$—the probability of the first row having a count of n and the second row having a count of m—is given by

$$P(n,m) = P(n) \times P(m)$$

where $P(n)$ and $P(m)$ are the probabilities of having a count of n or m, respectively. For example,

$$P(31,31) = P(31) \times P(31) = \left(\frac{13}{85}\right)\left(\frac{13}{85}\right) = \frac{169}{7225}$$

$$P(32,38) = P(32) \times P(38) = \left(\frac{12}{85}\right)\left(\frac{6}{85}\right) = \frac{72}{7225}$$

Rather than list all the possible probabilities, we find it more convenient to tabulate (see Table A.2) how many equally likely combinations, out of 7225 possible, lead to the various *coups*.

We see that there are 169 *refaits* out of 7225 *coups*; that there are $16 + 25 + 36 + 49 + 64 + 81 + 100 + 121 + 144$, or 636 *coups nuls*; and that therefore there are $7225 - 636 - 169$, or 6420 *coups*, in half of which the count of the first row is lower than the count of the second row and in the other half of which the reverse obtains. Since nothing happens in case of a *coup nul*, we may assume such coups to be nonexistent and consider only the 6589 valid *coups* ($7225 - 636$) as the possible events. Hence,

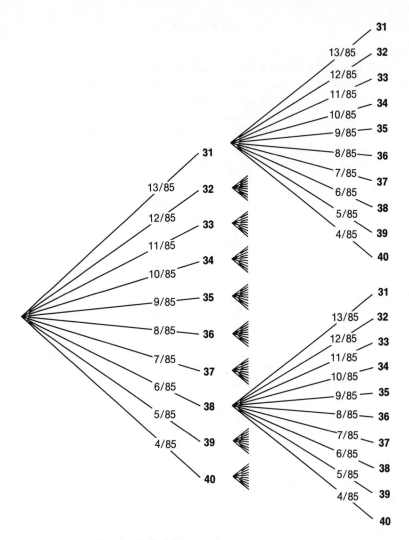

Fig. A.4 Probability tree for trente-et-quarante

**Table A.2**

| Second row | First row | | | | | | | | | |
| | 31 | 32 | 33 | 34 | 35 | 36 | 37 | 38 | 39 | 40 |
|---|---|---|---|---|---|---|---|---|---|---|
| 31 | 169 | 156 | 143 | 130 | 117 | 104 | 91 | 78 | 65 | 52 |
| 32 | 156 | 144 | 132 | 120 | 108 | 96 | 84 | 72 | 60 | 48 |
| 33 | 143 | 132 | 121 | 110 | 99 | 88 | 77 | 66 | 55 | 44 |
| 34 | 130 | 120 | 110 | 100 | 90 | 80 | 70 | 60 | 50 | 40 |
| 35 | 117 | 108 | 99 | 90 | 81 | 72 | 63 | 54 | 45 | 36 |
| 36 | 104 | 96 | 88 | 80 | 72 | 64 | 56 | 48 | 40 | 32 |
| 37 | 91 | 84 | 77 | 70 | 63 | 56 | 49 | 42 | 35 | 28 |
| 38 | 78 | 72 | 66 | 60 | 54 | 48 | 42 | 36 | 30 | 24 |
| 39 | 65 | 60 | 55 | 50 | 45 | 40 | 35 | 30 | 25 | 20 |
| 40 | 52 | 48 | 44 | 40 | 36 | 32 | 28 | 24 | 20 | 16 |

$$\text{Probability of } refait = \frac{169}{6589} = 0.0256488$$

$$\text{Probability of } noir = \text{probability of } rouge = \frac{3210}{6589} = 0.48755$$

Since the first card in the first row is just as likely to be red as it is to be black, it follows that the probabilities for *couleur* and *inverse* are the same as for *noir* and *rouge*.

## 6. ABSORBING MARKOV CHAINS

Let us back two *chances simples* (*à cheval*) at the European roulette table. Specifically, let us back *rouge-impair*. There are ten odd numbers that are red [see Fig. 2.3(a)]. Hence,

$$\text{Probability of } rouge\text{-}impair = \frac{10}{37}$$

Hence, the probability of winning a *coup* is

$$w = \frac{10}{37}$$

There are also ten even numbers that are black. Hence,

$$\text{Probability of } noir\text{-}pair = \frac{10}{37}$$

Since we lose in case of a *noir-pair* or a *zéro* and the probability of *zéro* is 1/37, we obtain for the probability of losing

$$\ell = \frac{10}{37} + \frac{1}{37} = \frac{11}{37}$$

That leaves *noir-impair* and *rouge-pair*, which lead to a *coup neutre*. There are eight even numbers that are red and eight odd numbers that are black. Hence,

$$\text{Probability of a } coup \text{ } neutre = \frac{16}{37} = n$$

Suppose we let our stake on *rouge-impair* ride until we either win or lose. (You may not have a choice; inquire about the rules before you start gambling!) The probability tree of Fig. A.5 applies to this situation. It represents what is known in the trade as an *absorbing Markov chain* (Ref. [6], p. 214ff). One can show (by mathematical magic) that it ends with a win or loss after finitely many steps, that is, it can't go on forever.

Specifically, we find the probabilities of eventually losing or

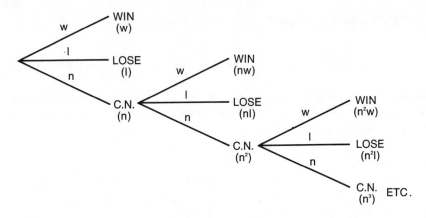

Fig. A.5 *Probability tree for* rouge-impair *in European roulette*

eventually winning from the tree probabilities in Fig. A.5, making
suitable use of Formula (1.2):

(6.1)    Probability of winning *rouge-impair* $= w + nw + n^2w + n^3w$

$$+ \ldots = w(1 + n + n^2 + n^3 + \ldots) = \frac{w}{1 - n} = \frac{10}{21} = 0.4761$$

(6.2)    Probability of losing *rouge-impair* $= \ell + n\ell + n^2\ell + n^3\ell + \ldots$

$$= \ell(1 + n + n^2 + n^3 + \ldots) = \frac{\ell}{1 - n} = \frac{11}{21} = 0.52371$$

(They add up to 1, of course, since $w + \ell + n = 1$ and hence, $1 - n$
$= w + \ell$.)

Since there are only nine odd numbers that are low and nine
even numbers that are high and 18 numbers that are even and low or
odd and high, the probability of winning, say *passe-pair*, is

$$w = \frac{9}{37}$$

the probability of a *coup-neutre* is

$$n = \frac{18}{37}$$

and we find from Formula (6.1) with these new values for w, n that the
probability of winning eventually is now

$$\frac{9}{19} = 0.4736842$$

which is slightly less than for *rouge-impair* (or *noir-pair*). So, if you must bet *à cheval* on two even chances, pick *rouge-impair* or *noir-pair*.

If we bet *à cheval* on two even chances in trente-et-quarante, we can find the various probabilities by using the same tree as in Fig. A.5 and the same formulas (6.1 and 6.2), but with different values for w, $\ell$, and n.

Suppose we back *noir-couleur*. Among 6589 valid *coups*, 6420 do not result in a *refait*. Assuming that *noir-couleur*, *noir-inverse*, *rouge-couleur*, and *rouge-inverse* are equally likely, we have

$$
\begin{aligned}
\text{Probability of } \textit{noir-couleur} &= \text{probability of } \textit{rouge-couleur} \\
&= \text{probability of } \textit{noir-inverse} \\
&= \text{probability of } \textit{rouge-inverse} \\
&= \left(\frac{1}{4}\right)\left(\frac{6420}{6589}\right) = \frac{1605}{6589} = w
\end{aligned}
$$

Since the probability of a *coup-neutre* is twice that much, or

$$n = 2w$$

we obtain from Formula (6.1) the probability of winning eventually:

$$\text{Probability of winning } \textit{noir-couleur} = \frac{w}{1 - 2w} = \frac{1605}{3379} = 0.4749926$$

This is slightly better than the probability of winning *passe-pair* (or *impair-manque*) at roulette but not quite as good as the probability of winning *rouge-impair* (or *noir-pair*).

## 7. EXPECTED VALUE

Suppose that you play some game in which you win an amount U with probability p, an amount V with probability q, an amount W with probability r, etc. Then,

$$Up + Vq + Wr + \ldots$$

is called the *expected value* of the game. In this formula, losses enter as negative winnings.

For example, if you back a number straight up *(en plein)* with a $1 chip at the American roulette table, where the pay-off is 35:1, the expected value of the game is

$$35\left(\frac{1}{38}\right) - 1\left(\frac{37}{38}\right) = -\frac{2}{38} = -0.526315$$

since you win $35 with probability 1/38 and lose $1 (win −$1) with probability 37/38.

If you back *rouge* at the European roulette table and do not take the *en-prison* option but rather forfeit half your stake after a *zéro*, then, on a one louis bet, the expected value of the game is

$$1\left(\frac{18}{37}\right) - 1\left(\frac{18}{37}\right) - \frac{1}{2}\left(\frac{1}{37}\right) = -\frac{1}{74} = -0.0135135$$

If you take, on the other hand, the *en-prison* option (with the understanding that only prisons of one degree are allowed), then you obtain instead

$$1\left(\frac{18}{37}\right) - 1\left(\frac{18}{37}\right) + 0\left(\frac{18}{37^2}\right) - \frac{19}{1369} = -0.0138788$$

as the expected value of the game, no matter whether it takes one spin or two spins of the wheel to bring it to a conclusion.

If the expected value of a game is positive, the game is favorable; if negative, the game is not favorable. If it is zero, the game is fair. No casino game (with the possible exception of blackjack) has nonnegative expected value. The casinos are not in it just for the fun of it; they want to make a profit!

If the expected value of a game is, say, −0.0135135 (as in backing a *chance simple* at the European roulette table and forfeiting half your stake if the zero comes up), then you may expect to lose in a large number N of such games, betting one monetary unit each time, an amount of 0.0135135N monetary units. If N is, for example, 1000, and if the monetary unit is one dollar, you may expect to lose $13.51, or thereabouts. By the same token, you may expect to lose with the simple prison option $13.88, etc.

## 8. HOUSE TAKE

A negative expected value of a game represents an expected loss for the gambler and, *eo ipso*, an expected gain for the gaming establishment. Since the expected value of betting one dollar straight up at the American roulette table is −0.0526315, the house may expect to gain from 10,000 such bets about $526.31—the *house take*.

When computing the house take from N games with expected value −e lasting one *coup* each, one simply multiplies the expected value, taken positively, by N, and that's that (house take per *coup* times number of *coups*). There are also games that may last longer than one *coup*, however, such as bets on *chances simples* at European roulette, trente-et-quarante with *prison* options, or *à cheval* bets on *chances simples*. For the house to obtain a meaningful measure for the profitability of a game, it has to compute the take relative to its effort— such as the number of valid deals or the number of spins of the wheel.

**Table A.3**

| Chance | Trente-et-quarante | | European roulette | | American roulette | |
|---|---|---|---|---|---|---|
| | Expected loss[1] | House take[2] | Expected loss[1] | House take[2] | Expected loss[1] | House take[3] |
| Even chance: | | | | | | |
| 0-prison option | $12.82 | $1,123.41 | $13.51 | $1,183.81 | $52.63 | $18,442.11 |
| 1-prison option | $13.15 | $1,123.41 | $13.88 | $1,183.81 | — | — |
| 2-prison option | $12.99 | $1,108.84 | $13.70 | $1,167.59 | — | — |
| 3-prison option | $12.99 | $1,108.58 | $13.70 | $1,167.35 | — | — |
| A cheval on even chance: | | | | | | |
| rouge-impair[4] vs. noir pair | $50.01 | $2,246.83 | $47.62 | $2,367.57 | — | — |
| passe-pair[4] vs. manque-impair | $50.01 | $2,246.83 | $52.63 | $2,367.57 | — | — |
| First five | — | — | — | — | $78.95 | $27,663.16 |
| All others | — | — | $27.03 | $2,367.57 | $52.63 | $18,442.11 |

[1] Per 1000 one-dollar bets that are left on the table until a final decision has fallen.

[2] For 87,600 spins of the wheel (or 87,600 valid deals in trente-et-quarante), which is equivalent to one spin (deal) every two minutes, eight hours a day, 365 days a year.

[3] For 350,400 spins of the wheel, which is equivalent to one spin every half minute, eight hours a day, 365 days a year.

[4] For trente-et-quarante, read: rouge-couleur vs. noir-inverse or rouge-inverse vs. noir-couleur.

To obtain such a measure, one has to compute the house take for a number of games and then divide by the number of spins it took to complete these games. It would lead us too far afield to go into any details. Suffice it here to say that it can be done via the computation of the expected number of *coups* (see Ref. [9]). We have made these computations for all possible bets at trente-et-quarante, at European roulette, and at American roulette and tabulated the results in Table A.3. To make the results more meaningful, we have considered a hypothetical player who bets one dollar 1000 times in succession on the same chance with the understanding that he lets his dollar ride until a game has come to a definite conclusion like win, lose, or break even. We have also assumed that a European roulette wheel is spun every two minutes, eight hours a day, 365 days a year, and that an American roulette wheel is spun every half minute, eight hours a day, 365 days a year. That makes it 87,600 *coups* per year for the European wheel and 350,400 *coups* per year for the American wheel. We have also assumed that 87,600 valid deals are dealt per year at one trente-et-quarante table.

When looking at this tabulation, you may be horrified about the huge expected (eventual) loss per game on an *à cheval* bet. Keep in mind that such a game lasts, on the average, almost two *coups*. Also observe that the expected loss on a bet on *rouge-impair* and *noir-pair* is less than on the other two combinations although it does not make any difference to the house. This is so because it takes, on the average, fractionally longer for a game on *passe-pair* or *manque-impair* to come to a conclusion. Note also that we have labeled a bet on a *chance simple* with half-the-stake-down-the-drain on zero as the 0-prison option.

# All BASIC Dialects Are Alike, Right? Wrong!

All our programs are in HP 2000 ACCESS BASIC. The reader should not encounter many difficulties in translating our programs into any of the other major BASIC dialects. To facilitate such a task, let us briefly discuss some of the more common deviations. (For the more subtle ones, we refer the reader to "BASIC REVISITED, An Update to Interdialect Translatability of the Basic Programming Language" by Gerald L. Issacs, CONDUIT, 1976, P.O. Box 388, Iowa City, Iowa 52440.)

*FOR-NEXT LOOPS*

After the exit from the loop,

```
10   FOR K = 1 TO N
20   HAVE SOME PEANUTS
30   NEXT K
```

the value of the variable K, in HP 2000 ACCESS BASIC, is N+1. (In some other dialects, it is N.) We made use of this fact in safeguarding against faulty inputs that are put through a "recognition-loop." (See line 360, Fig. 1.5; line 840, Fig. 2.14; and line 820, Fig. 4.4.)

*COMPUTED GOTO STATEMENT*

In HP 2000 ACCESS BASIC,

GOTO (numerical expression) OF (list of line numbers)

causes the computer to evaluate the "numerical expression," round it off to the nearest integer N, and transfer control to the line the number of which is listed in the Nth place in the "list of line numbers." If there is no such number, then the computed GOTO statement is ignored, and control passes to the next line in the program.

In other BASIC dialects, the computed GOTO statement may have a different form. For example,

ON (numerical expression) GOTO (list of line numbers)

or

GOTO (list of line numbers) ON (numerical expression)

may be encountered. They do exactly the same thing.

## COMPUTED GOSUB STATEMENT

Similarly, the computed GOSUB statement, which, in HP 2000 ACCESS BASIC, has the form

GOSUB (numerical expression) OF (list of line numbers)

may also be encountered in the form

ON (numerical expression) GOSUB (list of line numbers)

or

GOSUB (list of line numbers) ON (numerical expression)

## DELIMETERS

In HP 2000 ACCESS BASIC, a semicolon at the end of a PRINT statement supresses a line feed and a carriage return. In some other dialects, a comma may serve the same purpose. In some dialects, an apostrophe is used instead of a quotation mark at both ends of a character string. It may also be necessary to separate a string constant and a value within the same PRINT statement with a comma or a semicolon, such as in PRINT "X=",X (in HP 2000 ACCESS BASIC, PRINT "X="X will do).

## CHARACTER STRINGS

The admissible length of string constants and string variables varies with the BASIC dialect. Some dialects do not have a sub-string capability, and some do not allow the comparison of string variables (= or <>) in their level I basic interpreters. In such cases, the reader is well advised to require that inputs be made in some numerical code rather than in the form of character strings.

## RANDOM NUMBERS

In HP 2000 ACCESS BASIC, RND(1) produces a (pseudo)random number between 0 and 1. Some pseudo-random-number generators are not very satisfactory. How to program your computer to produce sequences of pseudo-random numbers may be found in Ref. [8] (p. 186).

# Appendix C

# Classification of Bets in Ascending Order of Expected Losses[1]

| Game | Bet | Expected Loss Per $1000 Bet |
|---|---|---|
| Craps | Don't pass (don't come) laying double odds | $5.84 |
| Craps | Pass (come) taking double odds | $6.06 |
| Craps | Don't pass (don't come) laying single odds | $8.18 |
| Craps | Pass (come) taking single odds | $8.48 |
| Chemin-de-fer | | $11.60[2] |
| Trente-et-quarante | *Chance simple* | $12.82 |
| European roulette | *Chance simple* | $13.51 |
| Craps | Don't pass (don't come) | $13.64 |
| Craps | Pass (come) | $14.14 |
| Trente-et-quarante | *A cheval* | $25.65[3] |
| American roulette | Even chance with half the stake back on 0 and 00 | $26.32 |
| European roulette | *A cheval* | $27.03[3] |
| European roulette | *En plein* | $27.03 |
| European roulette | *En plein, à cheval* | $27.03 |
| European roulette | *Transversale* | $27.03 |
| European roulette | *Sixain* | $27.03 |
| European roulette | *Colonne* | $27.03 |
| European roulette | *Colonne, à cheval* | $27.03 |
| European roulette | *Douzaine* | $27.03 |
| European roulette | *Douzaine, à cheval* | $27.03 |
| European roulette | *Carré* | $27.03 |
| European roulette | *Quatre premiers* | $27.03 |
| European roulette | *A cheval* on *rouge-impair* or *noir-pair* | $47.62[4] |
| Trente-et-quarante | *A cheval* | $50.01[4] |
| European roulette | *A cheval* on *pass-pair* or *manque-impair* | $52.63[4] |
| American roulette | Even chance | $52.63 |
| American roulette | Straight up | $52.63 |
| American roulette | Split | $52.63 |
| American roulette | Down the street | $52.63 |
| American roulette | Sixline | $52.63 |

| Game | Bet | Expected Loss Per $1000 Bet |
|------|-----|------------------------------|
| American roulette | Column | $52.63 |
| American roulette | Split column | $52.63 |
| American roulette | Dozen | $52.63 |
| American roulette | Split dozen | $52.63 |
| American roulette | Corner bet | $52.63 |
| Craps | Field | $55.56 |
| American roulette | First five (0, 00, 1, 2, 3) | $78.94 |
| Craps | Big six, big eight | $90.90 |
| Craps | Hard way (6, 8) | $90.90 |
| Craps | Hard way (4, 10) | $111.11 |
| Craps | Any craps | $111.11 |
| Craps | 7 | $166.67 |
| Craps | 2 | $166.67 |
| Craps | 12 | $166.67 |
| Craps | 3 | $166.67 |
| Craps | 11 | $166.67 |

[1]Blackjack is not included in this tabulation. The player, if he plays right, is expected to win in the long run. If he doesn't play right, no telling what he may lose!

[2]If the player draws with probability 1/2 to a 5 count and the banker must stay on 6 if the player stays. In all other cases the banker follows Table 3.1(b) in Chap. 3.

[3]If the stake is withdrawn in case of a *coup neutre*. See also footnote 4.

[4]If the stake is not withdrawn in case of a coup neutre but left riding until a decision has been reached.

## Appendix D

# French-English Mini-Vocabulary

*Note:* (*v.*) = verb; (*m.n.*) = masculine noun; (*f.n.*) = feminine noun; (*adj.*) = adjective; (*adv.*) = adverb; (*prep.*) = preposition; (*pl.*) = plural; (*def. art.*) = definite article; (*pron.*) = pronoun; (*p.p.*) = past participle

aller (*v.*) to go (va = it goes; ne va plus = it goes no longer)
après (*prep.*) after
avec (*prep.*) with

banco (*m.n.*) bank
bon (*f.* bonne) (*adj.*) good
boule (*f.n.*) ball

cagnotte (*f.n.*) pool, money box
carré (*m.n.*) square
carte (*pl.* cartes) (*f.n.*) card
chance (*f.n.*) chance, luck
chef (*m.n.*) principal, chief
chemin-de-fer (*m.n.*) railroad
cheval (*m.n.*) horse (à cheval = astride)
cinq (*adj.*) five
colonne (*f.n.*) column
comment (*adv.*) how
couleur (*f.n.*) color
coup (*m.n.*) blow, turn
croupier (*m.n.*) rider on the rump; attendant at gaming table
cylindre (*m.n.*) cylinder, roller

de of, from
dernier (*f.* dernière) (*adj.*) last
déshonneur (*m.n.*) disgrace
deuxième (*adj.*) second
dieu (*m.n.*) God
dix (*adj.*) ten
douzaine (*f.n.*) dozen
du (*replaces* de le) (*prep.*) of the, from the

est (*v.*) he (she, it) is (*see* être)
et and
être (*v.*) to be

faire (*v.*) to make

gagner (*v.*) to win
garage (*m.n.*) garage, shunting
grand (*m.n.*) the big one

impair (*adj.*) odd
interdire (*v.*) to veto, to prohibit
inverse (*adj.*) inverted, contrary

jeu (*pl.* jeux) (*m.n.*) game

la (*f. def. art.*) the
le (*m. def. art.*) the
les (*pl. def. art.*) the

manque (*m.n.*) lack, want, deficiency
mille (*adj.*) thousand
mise (*f.n.*) bid, bet
mon (*m. poss. pron.*) my
monsieur (*pl.* messieurs) (*m.n.*) mister
monstre (*m.n.*) monster
montant (*m.n.*) escalator, progression
moyen (*f.* moyenne) (*adj.*) middle

neige (*f.n.*) snow
n'est ce pas? ain't it?
neutre (*adj.*) neutral
noir (*adj.*) black
nul (*f.* nulle) (*adj.*) no, void

orphelin (*m.n.*) orphan

pair (*adj.*) even
parole (*f.n.*) word
partie (*f.n.*) party, game, contest
passe (*f.n.*) passing, passage
passer (*v.*) to go, to pass away, to fade
perdre (*v.*) to lose
personnel (*m.n.*) personnel
petit (*m.n.*) the little one
plein (*adj.*) full
plus (*adj.*) more
pour (*prep.*) for
pourboire (*m.n.*) tip, gratuity
premier (*f.* première) (*adj.*) first

prison (*f.n.*) prison, jail
privé (*p.p. of* priver) deprived, private

quarante (*adj.*) forty
quatre (*adj.*) four

refaire (*v.*) to redo, to recover
rien (*m.n.*) nothing
rigoureusement (*adv.*) rigidly
rouge (*adj.*) red

sabot (*m.n.*) shoe
salle (*f.n.*) hall
sept (*adj.*) seven
simple (*adj.*) simple, plain
six (*adj.*) six
sixain (*f.n.*) sixline
soir (*m.n.*) evening
sont (*v.*) they are (*see* être)
suivi (*p.p. of* suivre) followed
sur (*prep.*) on, upon

table (*f.n.*) table
tailleur (*m.n.*) dealer, tailor
tiers (*f.* tierce) (*adj.*) third
transversal (*adj.*) transverse (rue transversale = cross street)
trente (*adj.*) thirty
trois (*adj.*) three

un (*adj.*) one

va (*v.*) *see* aller
vingt-neuf (*adj.*) twenty-nine
voisin (*m.n.*) neighbor
vous (*pron.*) you

zéro (*m.n.*) zero

*. . . und gruen des Lebens goldner Baum.*[1]

[1] " . . . and green alone Life's Golden Tree." (Conclusion of the quotation from Goethe's *Faust* at the beginning of the book)

# Bibliography

(1) Culbertson, Ely, *Culbertson's Card Games Complete with Official Rules*, The Greystone Press, New York, 1952.

(2) Epstein, Richard E., *The Theory of Gambling and Statistical Logic*, Academic Press, New York, 1967.

(3) Fielding, Xan, *The Money Spinner*, Little, Brown and Company, Boston, 1977.

(4) Fleming, Ian, *On Her Majesty's Secret Service*, a Signet Book, 1964.

(5) Goodman, Mike, *Win*, Holloway House Publishing Company, Los Angeles, 1971.

(6) Kemeny, John G., Snell, J. L. and Thompson, G. L., *Introduction to Finite Mathematics*, 3rd ed., Prentice-Hall, Inc., Englewood Cliffs, New Jersey, 1974.

(7) McQuaid, Clement, *Gambler's Digest*, DBI Books, Inc., Northfield, Illinois, 1971.

(8) Sagan, Hans and Meyer, Carl D., *Ten Easy Pieces*, Hayden Book Company, Rochelle Park, New Jersey, 1980.

(9) Sagan, Hans, "Markov Chains in Monte Carlo," *Mathematics Magazine*, vol. 54, 1981.

(10) Scarne, John, *Scarne on Cards*, a Signet Book, 1973.

(11) Spencer, Donald D., *Game Playing with BASIC*, Hayden Book Company, Rochelle Park, New Jersey, 1977.

(12) Squire, Norman, *How to Win at Roulette*, GBC Books, Las Vegas, 1968.

(13) Thorp, Edward, *Beat the Dealer*, Vintage Books, New York, 1966.

(14) Wilson, Allan, N., *The Casino Gambler's Guide*, Harper and Row, New York, 1970.

# index